Mastitis: How t
clinical mastitis

A practical and easy to use guide to mastitis: 2

By Peter Edmondson

About the author

Peter Edmondson is a dairy veterinarian who has been specialising in mastitis and milk quality work for the past 35 years. He qualified from Trinity College Dublin in Ireland in 1980. After five years in practice in Ireland, Peter joined Almarai working with large dairy herds in Saudi Arabia and China. From there he was a partner in a large specialist dairy practice in the South West of England.

Peter formed UdderWise in 2015, a company which provides mastitis solutions. UdderWise specialises in providing mastitis technical expertise, milk quality programmes, tailor made training, troubleshooting mastitis, residue and milk quality problems, residue avoidance advice and providing legal expert witness work in all cattle and milk quality related matters.

Peter works for dairy farmers, vets, pharmaceutical, agricultural and dairy processing businesses throughout the world. He has excellent communication skills and is in demand to speak at international conferences.

He is a popular trainer of vets, farmers, technical and sales staff and is renowned for his practical and down to earth approach. He has been running intensive practical hands-on mastitis seminars for dairy vets and others from the dairy industry for over 20 years. Peter carries out referral visits throughout the world using his practical problem solving skills.

Peter is a Fellow of the Royal College of Veterinary Surgeons and a Registered Specialist in Cattle Health and Production.

He is a Diplomat of the European College of Bovine Herd Medicine.

Peter is involved in transferring skills and technology to help the dairy industry in the developing world. He spends time carrying out voluntary work in Africa.

Peter lives in Somerset in the South West of England with his family and a range of animals including dogs, cats, sheep, chickens and horses.

Peter can be contacted at:
peter@udderwise.co.uk

Bibliographic details

British Library Cataloguing in Publication Data
Mastitis: How to Control Clinical Mastitis
A practical and easy to use guide: 2

I. Edmondson, P.
ISBN 9781899043729

For more information on content/references please contact the author.

Context Products Ltd. 53 Mill Street, Packington, Ashby de la Zouch, Leicestershire. LE65 1WN England, Tel: +44 (0) 1530 411337
Email: admin@contextbookshop.com
www.ContextBookshop.com

About the Book
This is the second mastitis handbook to offer practical advice to help control clinical mastitis. The aim is to give the reader a thorough understanding of the bacteria that cause clinical mastitis, how these enter the udder and how to prevent infections occurring. The key focus is on practical solutions and there is an in depth explanation of the milking routine, management of the environment, dry period infections and other factors that can influence clinical mastitis. The use of mastitis records to help quantify mastitis and also pinpoint possible causes are described in detail, as are the economic consequences. This book is filled with photographs, figures and diagrams to give a clear explanation of the key points.

This book is a follow-up to the first handbook on how to control somatic cell counts. The cell count book follows a similar format giving an in-depth understanding of what cell counts are and why they are important, the bacteria responsible for high cell counts and how a herd goes about reducing cell counts to target levels. The book covers how to stop the spread of infection and how to manage high cell count cows. Both books have top tip sections summarising the key points to successful mastitis management. Both books are intended for farmers, herdsman, vets and anyone else who has an interest in mastitis control.

About Context
Context is a specialist publisher based in England producing targeted practical technical publications. Our expertise allows us to plan, design and produce publications that are colourful and easy to read yet providing you with the essential information you need quickly. We can also offer versions in different languages if required.

Sections

Introduction

How to control clinical mastitis

1

How to control clinical mastitis

Introduction

Clinical mastitis is the most common disease of dairy cows throughout the world. It has great welfare and economic significance and can be the cause of great stress to farmers.

Mastitis means inflammation of the mammary gland. Clinical mastitis causes visible changes to the milk such as clots, discolouration or a watery secretion. The infected quarter might be swollen and in severe cases, the cow will be ill. The majority of cases of clinical mastitis are mild.

FACTS

- Clinical mastitis is the most common disease in dairy cows
- Has great welfare and economic significance
- The majority of cases are mild
- Mastitis bacteria enter through the teat canal
- The bacteria do not pass from one quarter to another

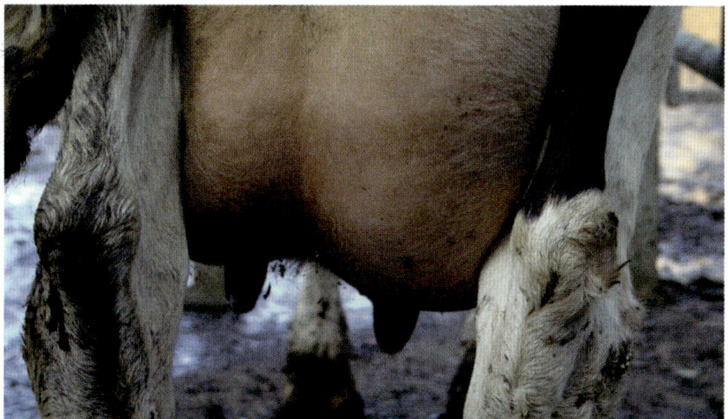

This cow has clinical mastitis in the back right quarter. You can see that it is much more swollen than the left back quarter

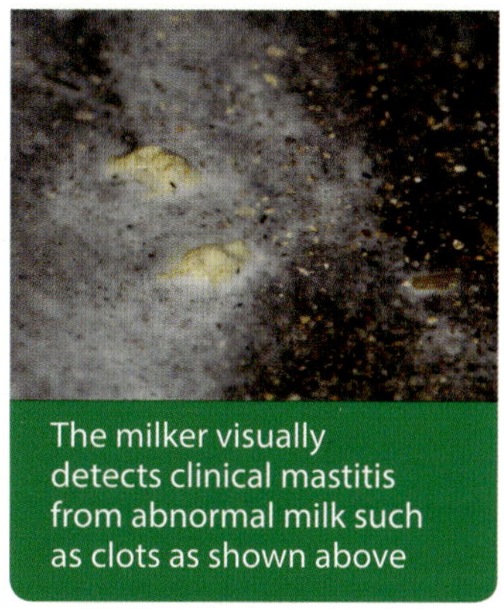

The milker visually detects clinical mastitis from abnormal milk such as clots as shown above

Mastitis is caused by bacteria which enter through the teat canal. The teat canal is between 6 and 10mm long. This is the only way that infection enters the udder. Mastitis infections do not pass from quarter to quarter. They are all totally separate and walled off from each other and have their own independent blood supply. Mastitis does come from animals drinking contaminated water or from the air. All bacteria have to physically enter the udder through the teat end and make their way into the udder.

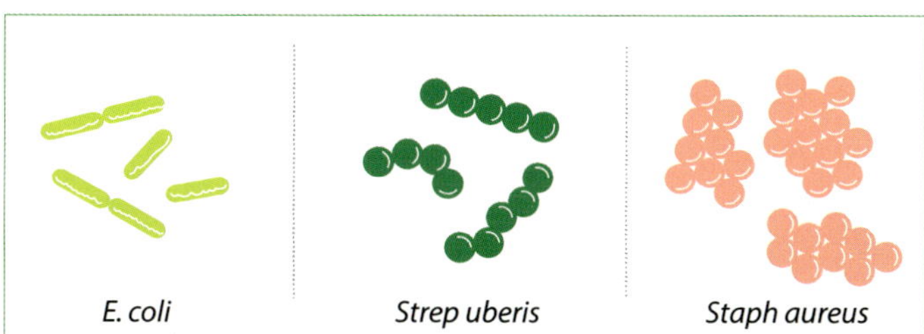

There are three key bacteria which cause clinical mastitis

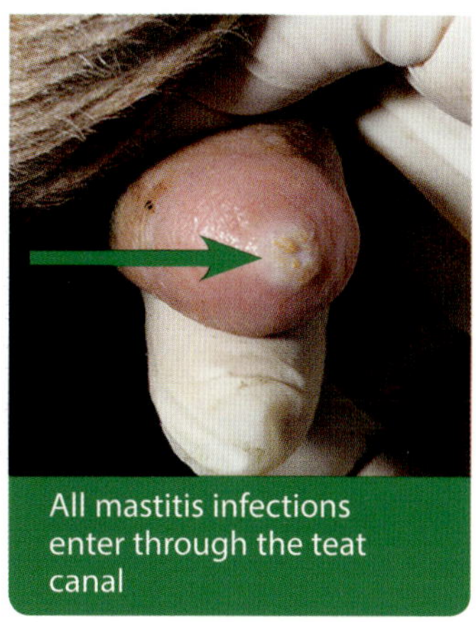

All mastitis infections enter through the teat canal

How to control clinical mastitis

Introduction

When any infection enters the udder, the cow's immune system tries to eliminate these bacteria. In many cases this is successful and the bacteria are eliminated. Mastitis occurs when the immune system is overwhelmed and unable to remove infections.

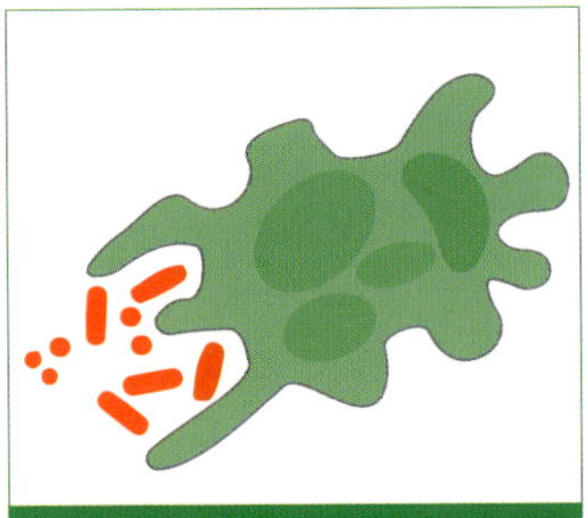

White blood cells trap bacteria and wall them off

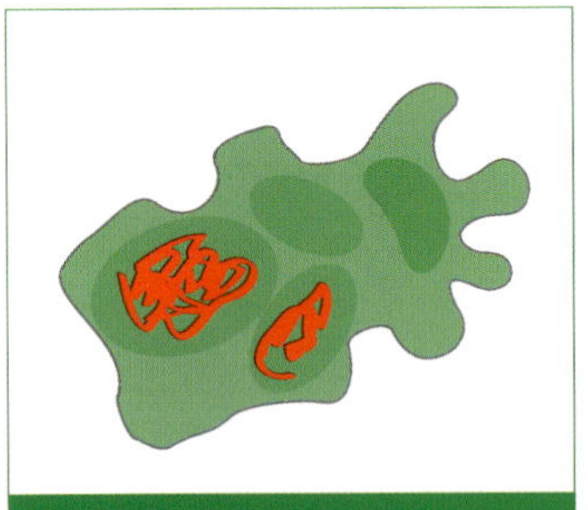

Enzymes then enter and digest the bacteria

FACT

- Mastitis occurs when the immune system is unable to eliminate infection

This shows the teat canal lined with Keratin, which is white. The teat canal is responsible for keeping mastitis bacteria out from the udder

How to control clinical mastitis

Environmental infections

The majority of bacteria which cause clinical mastitis come from the environment. The risk of mastitis increases when there is dirt or faecal matter coming into contact with the teat-end. *Escherichia coli*, or *E. coli* for short, is a typical example of an environmental bacteria. *Staphylococcus aureus*, commonly called *Staph aureus*, is another bacteria which causes clinical mastitis but lives in the udder and on the skin of the teats. *Staph aureus* does not live in the environment. It is a contagious mastitis bacteria; these live in the udder and on teat skin and are spread from cow to cow at milking time.

The majority of clinical mastitis comes from environmental bacteria. The risk of clinical mastitis increases with dirty teats

Mastitis can occur at all stages of lactation. The majority of clinical mastitis occurs in the first 100 days of lactation which coincides with peak milk production. Towards the end of lactation, the incidence of clinical mastitis reduces significantly.

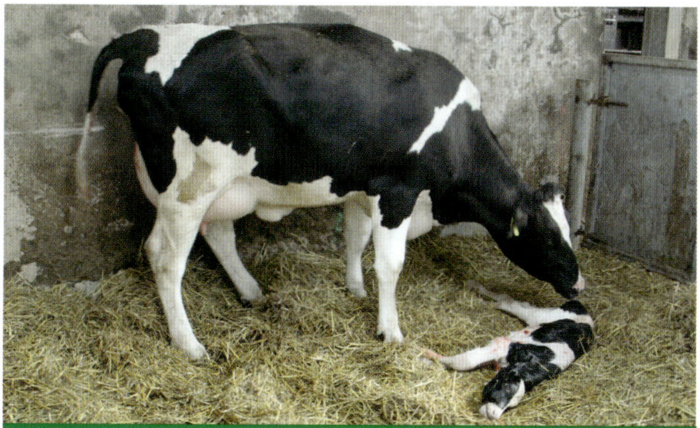

The freshly calved cow is at great risk of clinical mastitis

FACTS

- Environmental infections are the most common cause of clinical mastitis
- Dirty teats will increase the risk of clinical mastitis
- High-risk mastitis periods are early lactation and the beginning and end of the dry period

TOP TIP

- It is essential to identify the cause of mastitis in any herd

How to control clinical mastitis

Lactation and dry period infections

The majority of mastitis bacteria enter the udder during lactation, and are called lactation period infections or LPI for short. Some bacteria enter during the dry period and these are called dry period infections or DPI for short. If there are problems with dry period infections then you get more mastitis in the first 100 days of lactation.

It is important that each herd identifies which bacteria are responsible for clinical mastitis and at what stage of lactation they enter the udder.

Treatment success with clinical mastitis varies depending on the bacteria. The average cure rate is between 60 and 70%. Many farmers are frustrated by poor cure rates. Some cases recur in the same quarter, e.g. BL quarter 1st January, BL quarter 20th February. Recurrent or repeat cases are common with *Staph aureus* and *Strep uberis* clinical mastitis.

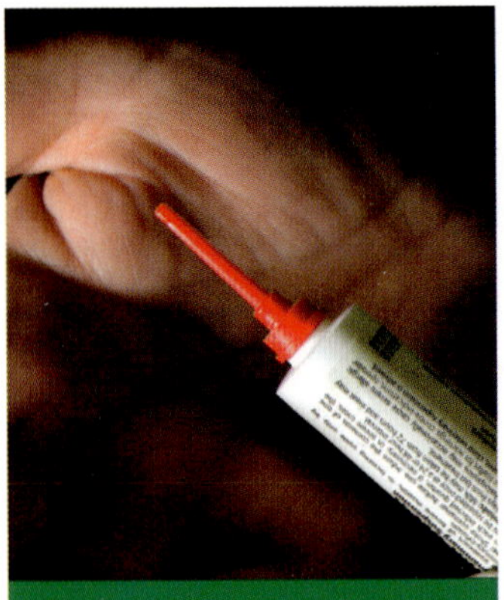

Treatment success for mastitis can be highly variable and will depend on the bacteria

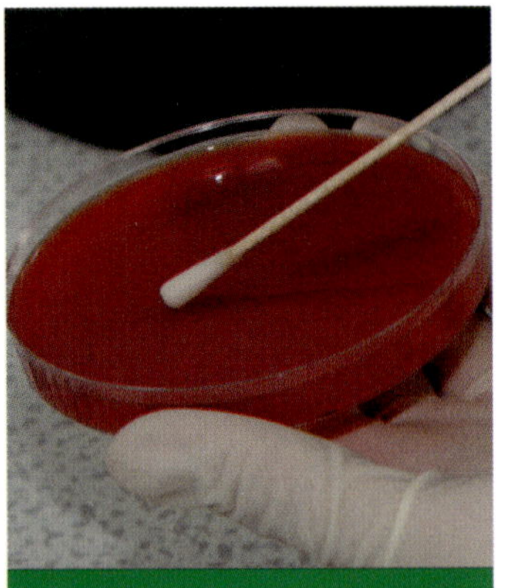

It is essential to know which bacteria are responsible for mastitis in your herd

The aim is to prevent mastitis occurring rather than relying on treatment. You cannot prevent all cases of clinical mastitis. No matter how hard you try there will always be times when faecal contamination gets onto the teat end and enters the udder. However, many herds have very low levels of clinical mastitis due to good mastitis management.

FACTS

- Dry period infections cause clinical mastitis in the first hundred days of lactation
- The majority of clinical mastitis occurs in the first hundred days of lactation
- The average treatment success for clinical mastitis is 60-70%
- Mastitis cure rates depend on the bacteria causing the infection
- You cannot prevent all cases of clinical mastitis occurring
- Many herds have very low levels of clinical mastitis

How to control clinical mastitis

Cell counts

Herd cell counts have reduced significantly in the UK and elsewhere over the past couple of decades due to improved mastitis management. This was encouraged by dairy companies changing their pricing structure favouring low cell count milk. As a result, the incidence of *Staph aureus*, a common cause of high cell counts, has reduced considerably. At the same time herds have increased in size and milk production.

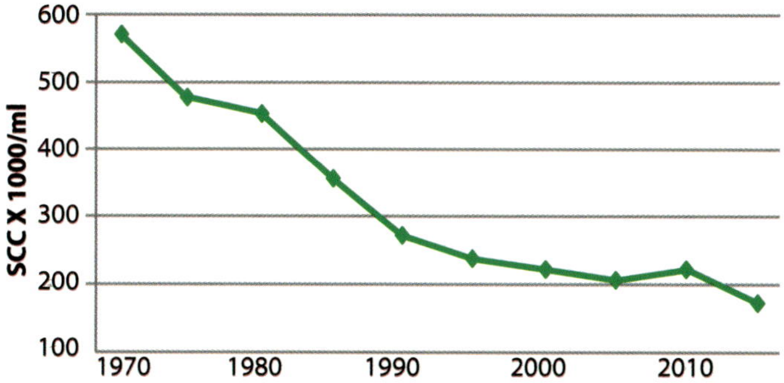

Changes in the average herd cell counts in England and Wales from 1970 to 2015

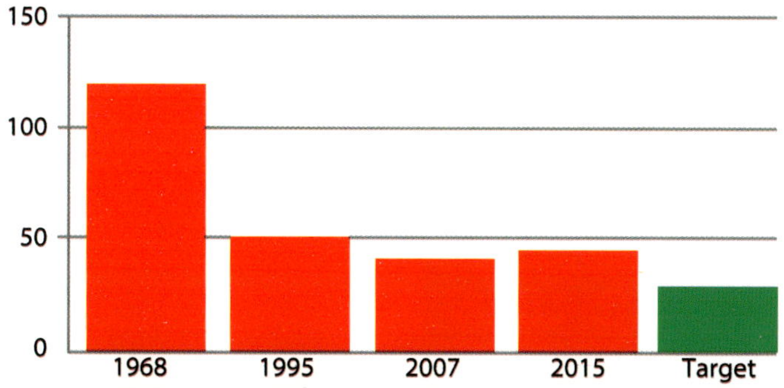

Average UK mastitis rate (cases per 100 cows/year)

In 1968 in the UK, there were 121 cases of mastitis per 100 cows per year compared to 47 in 2007. The mastitis incidence has more than halved over the last 30 years and in 2015 averaged between 40 and 50.

The target is to have 30 or less cases of clinical mastitis per hundred cows per year where one case is one quarter infected once. So a cow calving down with mastitis in four quarters will count as four cases of clinical mastitis.

FACTS

- Clinical mastitis levels have reduced by 60% in the last 50 years
- There is a massive range of clinical mastitis levels between herds
- A case of mastitis is one quarter infected once
- The target is 30 or less cases per cows per year

Changes to clinical mastitis

Cell counts

The greatest change in clinical mastitis over the past 40 years has been the decrease in *Staph aureus* infections. This bacteria was responsible for the majority of clinical mastitis in the 1960's and 70's and was the main cause of the high cell count at the time. Relatively few cases were caused by *E. coli*.

The graph below shows that in 1968 there was an average of 45 cases of *Staph aureus* for every 100 cows. This reduced to fewer than 8 in 1995 and down to less than 2 in 2007. Over the same time the incidence of *E. coli* has doubled due to increased environmental pressures combined with higher yielding cows.

FACT

- *Staph aureus* clinical mastitis levels of have reduced by over 80% in the last 50 years

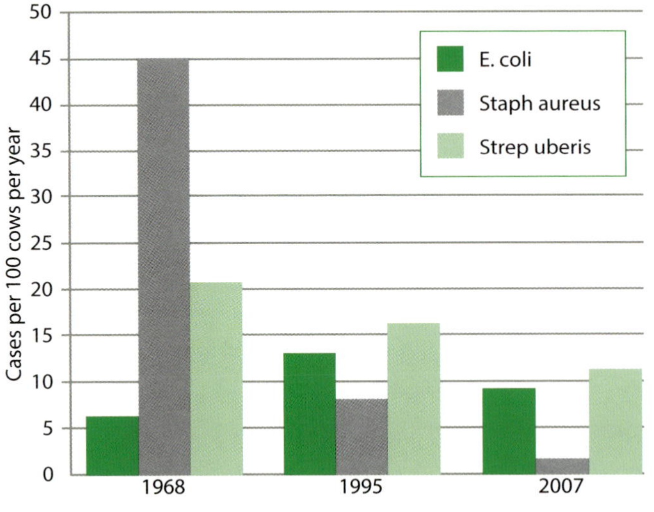

Changes to the incidence of mastitis bacteria over the past 50 years

Assessment of teat preparation in individual herds needs to be made in conjunction with the levels and types of clinical mastitis. For example if a herd has a low level of clinical mastitis then it would be unwise to make any changes to teat preparation, irrespective of how the teats are prepared, provided Bactoscan (a measure of bacterial numbers in milk) or TBC (total bacteria count) are low. If a herd has a high level of *aureus* clinical mastitis associated with poor teat preparation then changes must be made.

Changes to clinical mastitis

Milk flow rates

As you select for increased milk production you select for an increased risk in clinical mastitis. There is a positive correlation between faster milking cows and clinical mastitis. Changes to milk flow rate are shown to have doubled between 1950 and 1990 and this has resulted in a ten-fold increase in susceptibility of mastitis. This is due to higher yielding cows, improvements to the milking routine and milking technology.

FACT

- Faster milking cows are more prone to clinical mastitis

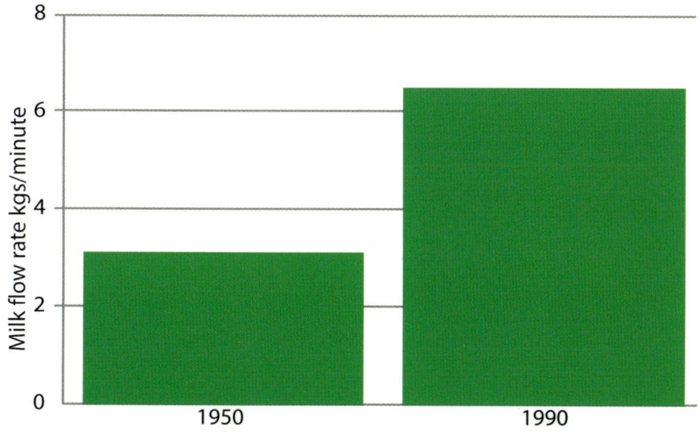

Changed to milk flow rate

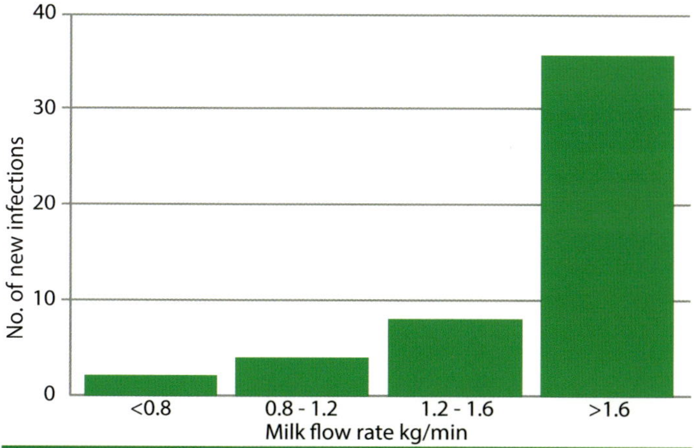

New infections and milk flow rates (quarter flow rate kg/minute)

Cows which milk out more quickly have more open teat canals. This allows bacteria easier access into the udder increasing the susceptibility to clinical mastitis as shown in the graph on the left.

However, improvements in milking routines and mastitis management have minimised the risks of more clinical mastitis cases occurring.

Changes to clinical mastitis

Environment

Many herds are now housing their cows at times when previously they would have been at pasture. Some herds have their cows housed all year round. This is very common with cows milked through robotic or voluntary milking systems (VMS). The risk of clinical mastitis increases when cows face more environmental challenge at the teat end.

Milking cows through an efficient and well maintained parlour helps to minimise mastitis

Many herds keep their cows housed all year round

As herds continue to expand in size and as yield increases, there is a potential for higher levels of mastitis. However, there are many herds which have improved their mastitis management and the cows environment so that udders and teats are clean.

This has resulted in;

- Lower levels of clinical mastitis
- Improved animal welfare
- Reduced antibiotic usage
- Increased production
- Satisfaction from a job well done

Well designed new buildings will keep cows clean and help minimise mastitis

Implications of clinical mastitis

Stress and worry

Dairy farmers are busy people and have considerable demands on their time every day. Farmers with high levels of mastitis often get worried and stressed. This can rise with toxic cases or if a cow dies. The response to treatments might be poor. Farmers can struggle to know how best to tackle the problem, will be concerned on the impacts on their business and the financial effects of clinical mastitis.

The table below shows words that farmers associate with mastitis. The most commonly associated words are in the biggest print size and the least common are the smallest. This shows that misery is an emotion very commonly associated with mastitis.

FACT

- High levels of mastitis cause worry and stress

Clinical mastitis can be a cause of great stress and worry

Time

Farmers like to milk dairy cows but they do not want problems. They want minimal levels of mastitis as this disrupts milking. It takes a considerable amount of time to manage cows under treatment. The additional tasks include;

- Withholding milk from the bulk tank
- Milking treated last or separately
- Administering treatments
- Recording all treatments
- Avoiding residues entering the bulk milk

It takes a lot of time to manage a cow with clinical mastitis

Implications of clinical mastitis

Welfare

Mastitis is very painful. The cow ends up lying on an inflamed udder. The greater the level of inflammation, the greater the level of pain to the cow. Some mastitis cases result in severe illness and even death. The welfare implications of clinical mastitis are very great, even with what appears to be mild cases.

FACT

- Mastitis is an extremely painful disease

This cow has acute *E. coli* mastitis and is in great pain

Economics

Mastitis is expensive. Milk from cows with clinical mastitis has to be discarded if cows are treated with medicines that have a milk withdrawal period. This milk cannot be sold. Treatment costs can be considerable, especially for cows with severe cases of mastitis. There are times where veterinary visits will be needed. These can all add up to considerable losses. The costs of clinical mastitis are discussed in detail in the 'Economics' section of this book.

FACT

- Clinical mastitis is very expensive and can reduce farm profitability

Mastitis is one of the most costly diseases of dairy cows

Implications of clinical mastitis

Poor response to treatment

Many cases of mastitis respond poorly to treatment. The average bacteriological cure rate, where the bacteria causing mastitis have been totally eliminated, averages between 60-70%. So, if you treat five cows, two will still have infection left in the udder tissue at the end of treatment. These cows are more likely to have a repeat or recurrent case in the same quarter later in lactation. Others might develop a high cell count. This can be very frustrating and worrying.

FACTS

- Some cases of clinical mastitis respond poorly to treatment
- Some cases keep recurring in the same quarter

Reduction in milk production

Mastitis losses can be the equivalent to several days milk production

FACT

- A case of clinical mastitis reduces lactation yield by 5%

This is the big hidden cost of clinical mastitis. Inflammation and infection cause severe damage to the udder tissue. On average, there is a 5% loss in milk production for the whole lactation following one case of clinical mastitis. So a cow that would normally produce 8,000 litres would only be producing 7,600 litres if she had one case of mastitis in her lactation!

The level of production loss will depend on the severity of infection. The production loss following an acute *E. coli* case immediately after calving where the cow dies will be 100%; the total yield for the lactation, and so this might be 8,000 litres.

A cow with a mild case of mastitis towards the end of her lactation might only have a small drop. But the average production loss is 5% per case irrespective of the stage of infection when mastitis occurred.

Implications of clinical mastitis

Culling

Some clinical mastitis cows end up being culled. For example, a severe *E. coli* infection where the cow dries up following mastitis or a cow with a high number of clinical cases of mastitis which do not respond to treatment. Culling is very expensive and limits which other cows can be culled from the herd for voluntary reasons such as poor production.

FACT

- Most cows with chronic mastitis end up being culled
- Toxic cases of mastitis can kill cows

Death

Deaths from mastitis are uncommon but do occur. They are always remembered! The most common cause of death will be from an acute *E. coli* infection, often at or around calving time. Most of these cases will have been treated prior to the cow dying and so not only is there the loss of the cow, but also veterinary and medicine costs, and the amount of time that the farmer spends nursing the cow. It can be very depressing as it is often the best cows which die.

Cows with repeat cases of clinical mastitis are often culled

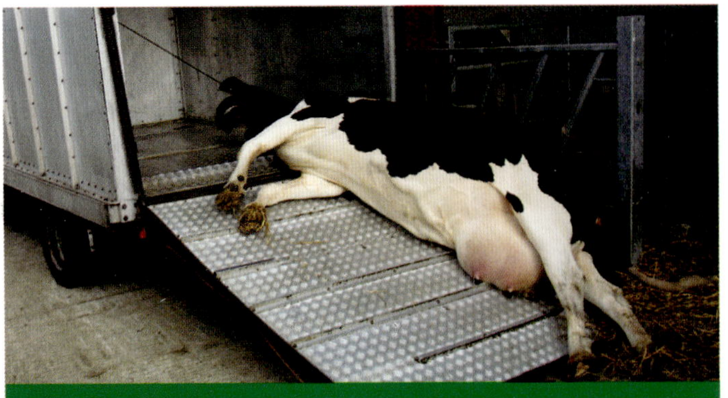

This cow died from *E. coli* mastitis

Farm assurance

All farms have to meet farm quality assurance standards. Many of these schemes are now focusing on disease incidence and a reduced use of antibiotics. Herds with high levels of mastitis have to have an action plan to reduce incidence. Mastitis and lameness are the two key diseases that assurance schemes focus on at present.

Implications of clinical mastitis

Antibiotic use

There is increased pressure to reduce antibiotic use in farm animals. Mastitis accounts for the majority of antibiotic use in dairy cows. Herds with a high incidence of clinical mastitis will have a high level of antibiotic use. High levels of clinical mastitis will also mean that more cows will need antibiotic dry cow therapy at the end of lactation.

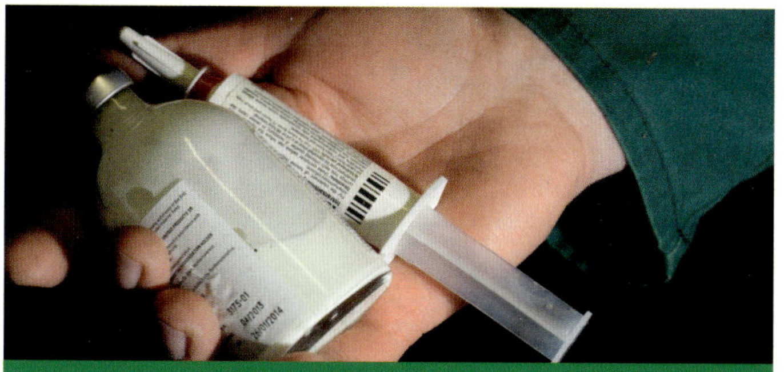

Clinical mastitis results in increased antibiotic use

FACTS

- There is considerable pressure to reduce antibiotic use in food producing animals
- Preventing disease reduces the need for antibiotic use

Fertility

Mastitis has an effect on fertility and this can result in;

- Shorter and less intense oestrus (heats)
- Reduced conception rates, especially around the time of clinical mastitis
- A longer interval from calving to conception
- An increased risk of abortion

In one trial, cows with clinical mastitis had a longer period to first service due to reduced oestrus behaviour. These cows needed more services per conception and the time from calving to conception was 25 days longer in cows that had clinical mastitis than healthy cows. Also note that high cell counts also affect fertility.

	Clinical mastitis	High SCC	Healthy
Days to 1st service	77.3	74.8	67.8
Days to conception	110	108	85
Services per conception	2.1	2.1	1.6

Economics

How to control clinical mastitis

2

Economics of clinical mastitis

Calculating the costs

Mastitis is one of the few diseases where costs can easily be calculated.

These include;

- Discarded milk during treatment
- Medicines
- Labour
- Veterinary fees
- Reduction in yield for the rest of lactation
- Deaths
- Culling and loss of genetic potential

FACTS

- The costs of mastitis are frequently underestimated
- The cost of medicines is low in comparison to the total cost of mastitis
- Clinical mastitis takes time to treat and manage

Discarded milk

The milk discarded is easy to work out. You add the treatment time to the withdrawal period and multiply by the average yield of the cow. For example, a cow giving 40 litres per day is treated with antibiotics for three days followed by a milk withdrawal period of four days.

The milk discarded for this case will be;

Discarded milk = 40 litres/day x 7 days (3 days treatment and 4 days milk withdrawal) = **280 litres**

Discarded milk

Medicines

This will include intramammary tubes together with any injectable antibiotics and other medicines. More herds are now using the hormone oxytocin to help ensure that mastitis cows are milked out fully. The use of non-steroidal anti-inflammatories (NSAID's) for pain relief and to speed up healing is more commonly used. Add on the costs of any other medicines used by you or your vet.

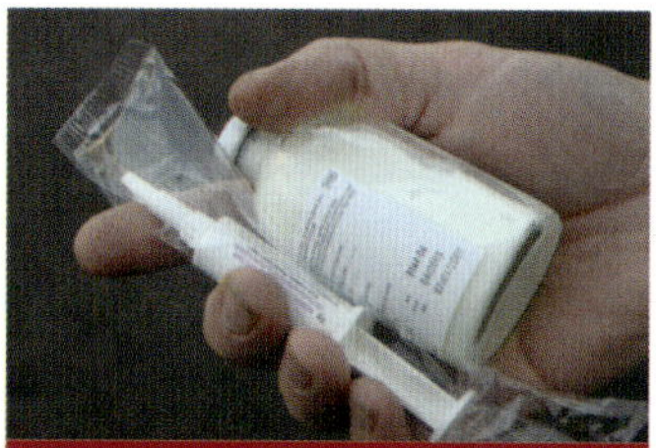

The costs of medicines are a small proportion of the total costs of mastitis

Economics of clinical mastitis

Calculating the costs

Labour

All farmers appreciate that a simple case of mastitis slows down milking; the cow has to be milked separately, medicines administered, records completed, time nursing cows and milk discarded for the withdrawal period.

Most farmers will tell you that this adds up to about an hour per case. For the more severe case, where animals need to be nursed and stripped out frequently, the labour input is significantly high.

FACT

- A case of clinical mastitis results in a 5% reduction in milk yield for the whole lactation

Veterinary costs

In some countries a vet has to visit and examine each cow before antibiotic treatment can be administered and so veterinary costs here will be higher. In other countries, a vet can dispense mastitis treatments to be left on farm.

Most farmers request their vet to attend cows with toxic mastitis. Many of these cows will need to be seen again. Good herds will require few visits as the severity and incidence of mastitis will be low.

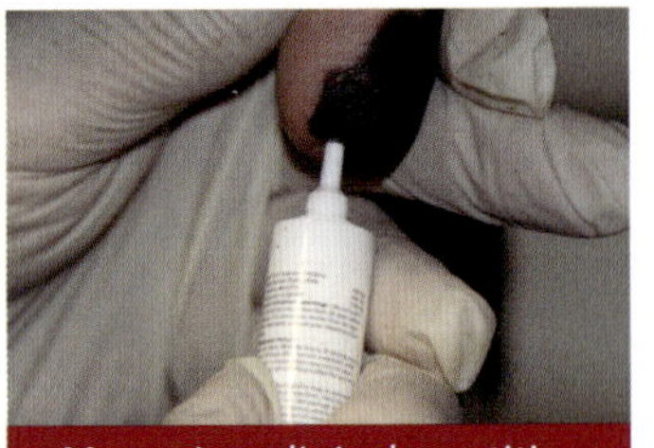

Managing clinical mastitis takes time

Production loss

The majority of people underestimate the production loss following a case of mastitis. Clinical mastitis causes severe damage to the milk producing tissue which means a reduction in production for the rest of lactation. Mild cases can result in a relatively small reduction whereas an acute *E. coli* case often causes the udder to dry up altogether resulting in a production loss for the entire lactation.

Many cows with mastitis are treated with NSAIDs

The average production loss is 5% of total milk yield following any case of clinical mastitis, irrespective of when the clinical case occurred. So the production loss for an 8,000 litre cow with one case of mastitis will be 400 litres (8,000 litres x 5%).

Take a herd averaging 8,000 litres with 50 mastitis cases each year. The loss of production for the herd will be;

Production loss = 8,000 litres/cow x 5% production loss/case = 400 litres/case x 50 cases **= 20,000 litres per year**

Economics of clinical mastitis

Calculating the costs

Deaths

The majority of mastitis deaths will be due to *E. coli* and occasionally gangrenous mastitis. Gangrenous mastitis can also cause death. The total cost of a death following clinical mastitis will be much greater than just the value of the animal that has died. It will also include loss of genetic potential, disposal costs plus the total loss of production and profit from the rest of the lactation, plus any treatment costs and time spent nursing the cow before death.

A dead cow is a significant loss for any dairy farmer

Culling

A herd with good mastitis management will have to cull very few, if any animals due to mastitis. Cows with repeat cases of clinical mastitis are more likely to be culled. The cost will be the difference between the sale and replacement value. To this should be added the difference in yield of the replacement animal.

Culling good cows due to mastitis problems is expensive and depressing

For example, a 10,000 litre cow in lactation four has been culled for £1,000. She has been replaced by a heifer which gives 7,000 litres and costs £2,000. The immediate cost of culling is therefore £1,000, the difference between the sale and replacement value. To this should be added the difference of 3,000 litres of milk lost for the first lactation, and subsequent differences in production that occur until she reaches the production of the cow that was culled. Often this production loss is higher than the difference in the cull and replacement value of the animal.

Cost of mastitis

Calculating the costs

You can use this simple table to give you an indication of the cost of mastitis in your herd. It does not cover all costs as does not include the costs of deaths, culling or veterinary fees which can be added on. It gives an indication of your basic clinical mastitis costs.

	Description	£	Your herd
Milk discarded	No. days treatment and No. days milk withheld = A Average yield of a cow with mastitis = B Milk price (in £, € or $ etc) = C	A x B x C	
Treatment costs	Average No. milking cow tubes x cost/tube = D Average cost of other treatments = E	D + E	
Labour	Time taken to manage a case of mastitis, add the hourly labour rate (F) x time taken per case in hours (G)	F x G	
Lost production	Average yield per lactation (H) x milk price (in £, € or $) (C) x 5%	H x C x 0.05	
Total		Add up the figures above	

Worked example

1. This herd averages 8,000 litres and the milk price is £0.30/litre
2. Cows are treated for three days with a four day milk withdrawal period
3. The average yield for cows with mastitis is 40 litres/day
4. Cows are given three intramammary tubes at £2/tube
5. Cows receive a non-steroidal anti-inflammatory treatment which costs £20
6. It takes an hour to manage a mastitis cow and labour costs are £10/hour

The average cost of a case of mastitis is £240 when all the figures are put into the table below. This does not take into account any vet fees, culling or deaths that might have occurred.

	Description	£
Milk discarded	3 days treatment + 4 days discard = 7 days x 40 litres x £0.3	84
Treatment cost	3 tubes x £2 + Non-sterodial injection @ £20	26
Labour	1 hour of labour x £10/hour	10
Lost production	8,000 litres x £0.3 x 5%	120
Average cost of clinical mastitis		**£240**

FACT

- The biggest cost of clinical mastitis is lost production due to udder damage

The pie chart shows the costs of mastitis for this example. This clearly shows that the greatest of cost is the production loss followed by the milk discarded. The costs of medicines are relatively insignificant.

Cost of mastitis

The impact of mastitis on profitability of three different herds

Description	A	B	C
Number of clinical cases	20	60	90
Average cost per clinical case	240	270	270
Cows culled @ £1,000 each	0	2	3
Cows died @ £2,000 each	0	0	1
Total cost of clinical mastitis	4,800	16,200	24,300
Cost of culling	0	2,000	3,000
Cost of deaths	0	0	2,000
Total costs of clinical mastitis	4,800	18,200	29,300
Losses as pence per litre	0.6 ppl	2.3 ppl	3.7 ppl
Equivalent reduction in milk price compared to Herd A		-1.7 ppl	-3.1ppl

Analysis

These three herds (A, B and C) all have 100 cows giving an average of 8,000 litres and have had different levels and severity of clinical mastitis over the past year.

Herd A is a very well managed herd with 20 cases of mastitis, well below target levels. The cases respond very well to three days of treatment.

Herd B has 60 mastitis cases. Treatment success is poor and most cows need 5 days treatment and this adds to the cost of treatment. There are problems with repeat cases and two cows were culled.

Herd C had 90 cases of mastitis, one cow died and three cows were culled for mastitis. Treatment success is poor and the costs per case are the same for Herd B. This farm has very poor mastitis management.

Effect on profit

Assume that each farm had a profit level of £30,000 per annum without taking into account the costs of clinical mastitis. The graph below clearly shows the difference in farm profitability related to clinical mastitis between the three herds. This does not take into account any cell count or Bactoscan penalties.

Most farmers do not work out their mastitis incidence or costs. This clearly shows that clinical mastitis can have a significant effect on profit in many herds, especially those which are poorly managed, are not aware of these losses.

Impact

Herd A is losing the least at £4,800 or £48/cow per year, equivalent of 0.6 pence per litre (a 2% reduction in milk price). You can't completely eliminate mastitis but it can be reduced to low levels as shown in this herd.

Herd B costs are very high at £18,200 or £182 per cow, equivalent to a milk price reduction of 2.3 pence per litre, an 8% reduction in milk price. This herd is losing 1.7 pence per litre more than Herd A for clinical mastitis costs.

Herd C will be struggling to make a profit. Mastitis is costing £29,300 per year or £293 per cow, equivalent to a reduction of 3.7 pence per litre, a 12% reduction in milk price. This herd is losing 3.1 pence per litre more than Herd A for clinical mastitis costs.

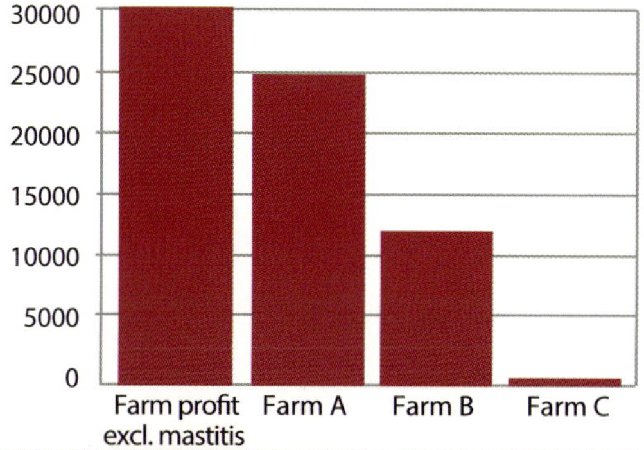

Farm profitability showing effect of clinical mastitis

Targets & Monitoring

How to control clinical mastitis

3

Record Keeping
Introduction

Mastitis is one of the few diseases where a detailed analysis of the data is extremely valuable. It can quantify the level of mastitis and to help identify the cause and source of infection. Records allow herds to compare mastitis levels against target figures and other herds. Records are an important part of monitoring the incidence of any disease.

Many farmers keep mastitis records but these are often not analysed and so the incidence of clinical mastitis is often underestimated. It is important to make good use of records otherwise there is little benefit to be gained from keeping the data.

Mastitis records will enable the farmer to do the following:

- Monitor the herd mastitis performance
- Compare incidence with other herds
- Identify the cause of clinical mastitis
- Identify cows whose milk needs to be withheld from the bulk supply
- Identify problem cows that should be considered for culling

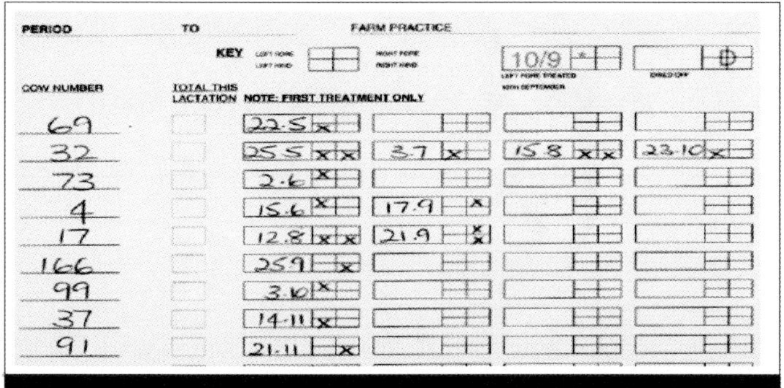

PERIOD TO FARM PRACTICE

KEY

COW NUMBER	TOTAL THIS LACTATION	NOTE: FIRST TREATMENT ONLY			
69		22.5			
32		25.5	3.7	15.8	22.10
73		2.6			
4		15.6	17.9		
17		12.8	21.9		
166		25.9			
99		3.10			
37		14.11			
91		21.11			

Manual record showing mastitis cases

One case of mastitis is defined as one quarter infected once. A cow that calves down with mastitis in all four quarters counts as four cases of mastitis.

Mastitis cases can be recorded in a variety of ways. With manual records, subsequent cases of mastitis should be recorded adjacent to the first so that problem cows are easily identified.

Using the system above, it is clear that cow 32 has had six cases of mastitis and four of these were in the back left (BL) quarter. Although exactly the same information is recorded on the following table (on page 22), it's not so clear that cow 32 is such a problem.

KEY POINTS

- Records allow herds to compare mastitis incidence
- Records can help to establish the cause of mastitis
- Most farmers underestimate clinical mastitis

TOP TIP

When recording cases of mastitis, keep records of the following;

- Cow number
- Date
- Quarter/s infected
- Severity of mastitis
- Treatments administered
- Bacteriology results
- Outcome

FACTS

- A case of mastitis is one quarter affected once
- Simple mastitis records can be very effective in monitoring the disease

Record keeping

Information and targets

Cow	Date	Quarter
69	22.5	BL
32	25.5	BL + BR
73	2.6	FL
32	3.7	BL
4	15.6	FL
17	12.8	BL + BR
32	15.8	BL + BR
4	17.9	FR
17	21.9	BR + FR
166	25.9	BR
99	3.10	FL
32	23.10	BL

Farm Records

Simple mastitis records can be very effective. The bar chart below shows that the target number of clinical cases is three per month. The cow number, quarter and date of each case is recorded in a single box. If the cow has mastitis in two quarters, then this is recorded as two boxes, for example cow 6 had mastitis in the left and right back quarters in January.

No of cases	Jan	Feb	Mar	Apr	May	Jun	Jul	Aug	Sep	Oct	Nov	Dec
10												
9												
8	6LB											
7	6RB		81RF									
6	4LF	8LB	81LF									
5	101RB	99RF	81RH									11LB
4	19LF	17LB	81LH	35RB			Monthly Target				82LB	LLRB
3	6LB	1LF	23RB	35LB						6LB	31RF	77LF
2	43RF	9LB	13LH	101LB	66RH		71RB	101RB	32LB	79LB	21RB	9RB
1	31LB	14RF	87LH	41RF	31LB	17LB	91LF	77LB	66RB	17LB	49LF	53LB

Key LB= Left back, LF= Left front, RB= Right Back, RF= Right front

Record keeping

Computerised record keeping

More herds now have computerised recording systems which allow varying abilities for analysis. Some, like the INTERHERD programme, have good analytical properties which are invaluable in disease monitoring and identifying causes of the mastitis problem.

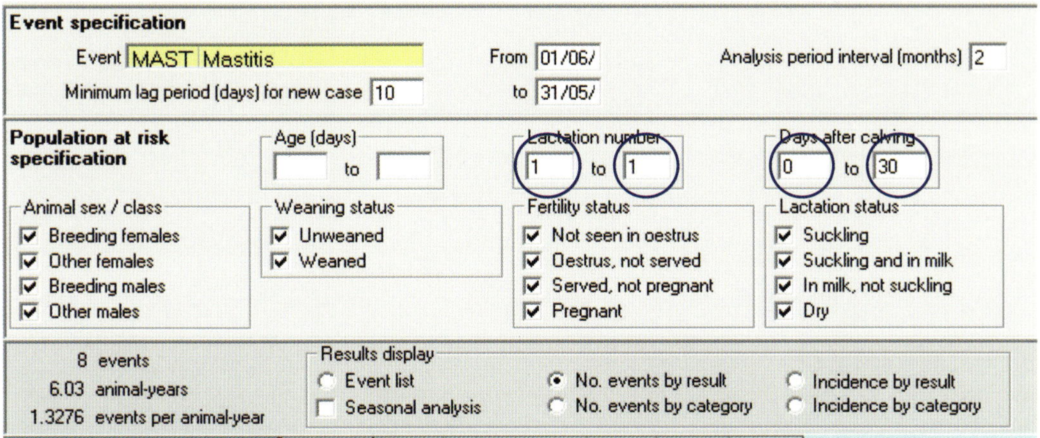

Result	Overall	01/06	01/08	01/10	01/12	01/02	01/04
Overall	8	0	2	1	4	1	0
A - Mild	8	0	2	1	4	1	0
B - Moderate	0	0	0	0	0	0	0
C - Severe	0	0	0	0	0	0	0

Computerised recording systems allow mastitis analysis to be carried out at the click of a button. Some programmes allow you to work out the incidence in a very detailed way ~ for example, the analysis above shows that there were 8 cases of mastitis in heifers in the first 30 days of lactation over a six month period, and four of these were in December and all cases were mild. Not all computer systems offer such analysis and some are very limited in terms of what analysis they can provide.

Mastitis parameters

Mastitis rate and percent herd affected

Mastitis rate

The mastitis rate is the number of cases of mastitis per 100 cows per annum, where one case is one quarter. It is a really useful indicator of mastitis incidence as it allows comparison between herds irrespective of size. The target is a mastitis rate of 30 or less.

A mastitis rate below target means that the herd has good control of clinical mastitis, assuming that all cases of mastitis have been recorded!

$$\text{Mastitis rate} = \frac{\text{No. of mastitis cases in a year} \times 100}{\text{Average no. of cows in herd for the year (milking and dry)}}$$

A high mastitis rate indicates a high incidence of mastitis.

Worked example

A herd of 150 cows has had 90 cases of mastitis in the last year.

$$\text{Mastitis rate} = \frac{90 \text{ (cases)} \times 100}{150 \text{ cows}} = \mathbf{60 \text{ cases/100 cows per year}}$$

The mastitis rate in this herd is twice the target level of 30

FACTS

- The mastitis rate is the number of cases of mastitis per 100 cows per annum
- The mastitis rate allows comparison between herds irrespective of size

TARGET

- Mastitis rate of 30 or less
- 20% of cows with mastitis per year

KEY POINT

- A case is one quarter with mastitis

Percentage herd affected

The percentage of cows affected per year represents the proportion of the herd that have had one or more cases of mastitis over a 12 month period. This helps to give some indication of the type of mastitis present. The target is 20% of cows with one or more cases of mastitis in a twelve month period.

Environmental mastitis is likely to result in a larger percentage of the herd affected as these infections are caused whenever cows come into contact with environmental contamination. This will typically occur in herds where the udders and teats are dirty and the risk of *E coli* and *Strep uberis* infections will be high. Herds with major problems with environmental mastitis will have 25% or more of the herd affected.

Chronic recurring mastitis, such as *Strep uberis* or *Staph aureus,* can affect a small proportion of cows, but there may still be a high mastitis rate. This can occur because the same cows keep getting recurrent cases of mastitis.

$$\text{\% cows affected} = \frac{\text{No. cows that had one or more cases of mastitis over a 12 month period} \times 100}{\text{Average no. of cows in herd for the year (milking and dry)}}$$

Worked example

60 cows had one or more cases of mastitis in a herd of 150 cows in the last year.

$$\text{\% cows affected} = \frac{60 \text{ (cows)} \times 100}{150 \text{ (total cows in herd)}} = \mathbf{40\% \text{ of cows affected with mastitis}}$$

40% of cows were affected with mastitis which is twice the 20% target figure. This indicates that environmental mastitis is a problem in this herd.

Mastitis parameters

Recurrence rate

A recurrent or repeat case of mastitis refers to one or more cases of mastitis occurring in the same quarter. If a treated quarter clears up and mastitis recurs seven or more days after the end of clinical signs, this is defined as a new case of mastitis. The target is a recurrence rate of 10% or less.

The recurrence rate is the percentage of quarters requiring one or more repeat treatments over a 12 month period.

It can be difficult to work out the recurrent quarters. Take the mastitis cases below where five cows have had mastitis. Cow 146 has had three cases of mastitis in the back right quarter. This counts as one quarter infected with mastitis and one recurrence. This means that for cow 146 the recurrence rate is 100%. It does not matter if the cow had 10 cases of mastitis in the back right quarter; it still only counts as one recurrent case.

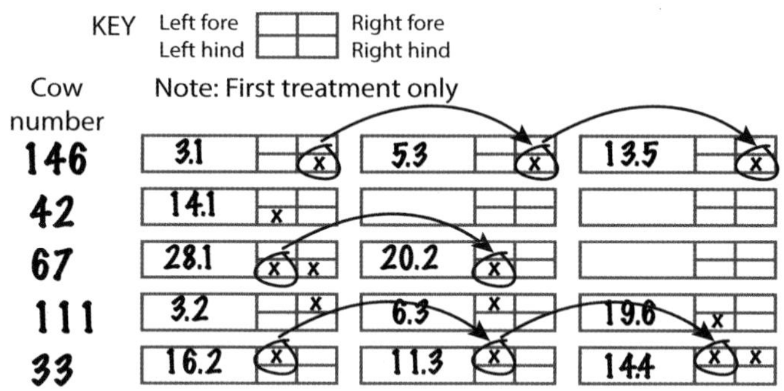

Look at the records for cow 33. She has had two quarters with mastitis, the front left and front right. There have been three cases in the front left quarter and so this counts as one recurrence. For this cow, the recurrence rate is 50% ~ one recurrent quarter out of two quarters with mastitis.

$$\text{Recurrence rate} = \frac{\text{No. quarters requiring one or more repeat treatments} \times 100}{\text{Total no. of quarters affected with mastitis}}$$

The five cows above had a total of 9 quarters with mastitis. Three of these quarters have had a recurrent case, and so the recurrence rate is 33% (3 repeat quarters out of nine with mastitis).

Worked example

There were 80 quarters affected with mastitis and 16 of these had two or more cases of mastitis.

$$\text{Recurrence rate} = \frac{16 \text{ (Total no. quarters requiring one or more repeat treatments)} \times 100}{80 \text{ (total no. of quarters affected with mastitis)}} = \mathbf{20\%}$$

The recurrence rate is 20% compared to the target of 10%. This suggests *Strep uberis* or *Staph aureus* mastitis, poor mastitis detection or a poor treatment regime.

TARGETS

- A recurrence rate of 10% or less

FACT

- High recurrence rate suggests problems with *Staph aureus* or *Strep uberis* mastitis, poor mastitis detection or treatment problems

A high recurrence rate may be due to problems with *Staph aureus* or *Strep uberis* infections which can be difficult to eliminate. High rates can also be due to poor mastitis detection where infections are not picked up early and so the response to treatment is poor. Likewise, if incorrect treatments are used, such as too short a duration or the incorrect selection of antibiotic, then cases can recur.

Mastitis parameters
Seasonal variation

It is useful to examine mastitis incidence by month. For example, a high number of cases in the housed period suggests a problem due to environmental mastitis. Likewise, a high number of cases in certain months might be related to a high number of calvings or a time when cows were dirty. It should be possible to minimise these effects with good management and facilities.

All year round mastitis incidence above target levels suggests contagious mastitis or constant environmental conditions. This would be common in herds where cows are housed all year round.

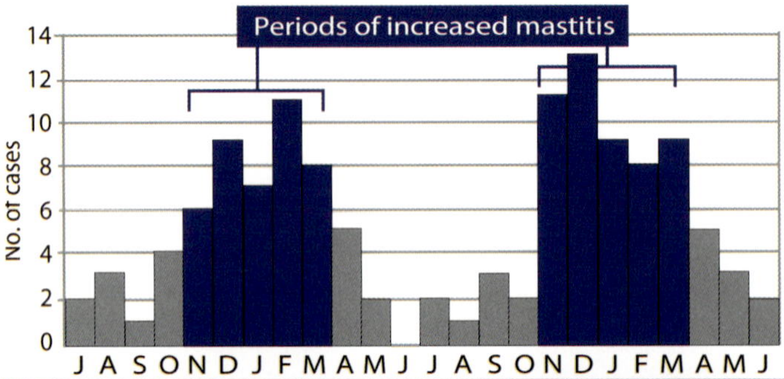

Graph showing the occurrence of mastitis per month in relationship to the housed period

This graph shows that the majority of mastitis is occurring during the housed period from November to March each year. This is typical of environmental mastitis due to poor housing conditions.

Cases within 30 days of calving

You can get an indication of infections arising from the dry period by assessing the mastitis incidence within 30 days of calving. The target is to have at most one cow with mastitis out of twelve (8%) within 30 days of calving. A high incidence suggests dry period infections are a problem. Likewise, a lower incidence suggests that the dry period is not contributing significantly to clinical mastitis. This is a very useful way of assessing the risk of dry period infections.

Worked example

12 cows had one or more cases of mastitis within the first 30 days of calving out of 80 cows that calved.

% cows with clinical mastitis within 30 days of calving = $\frac{\text{12 cows with mastitis}}{\text{80 cows that calved}}$ = **15%**

15% of cows had mastitis within 30 days of calving and this is twice the target figure. This suggests that dry period infections are a significant problem.

FACT

- A high number of cases in the housed period suggest a problem due to environmental mastitis
- Dry period infections are a problem if more than 8% of cows have mastitis within 30 days of calving

TARGET

- One cow out of 12 cows (8%) with mastitis in the first 30 days of lactation

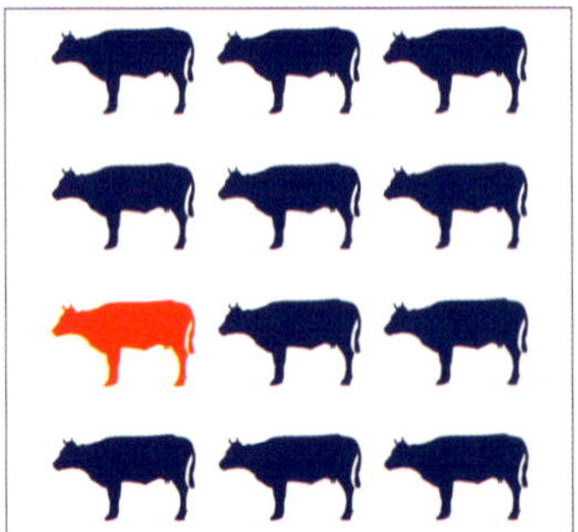

Cases in 1st 30 days target

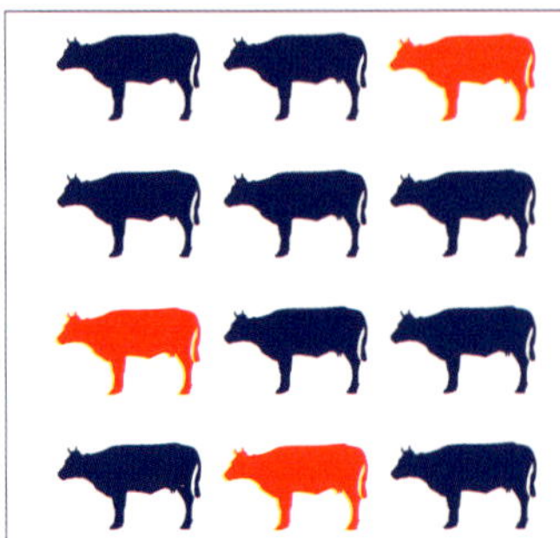

Cases in 1st 30 days problem

Mastitis parameters
Stage of lactation

Approximately 50% of mastitis cases occur within the first hundred days of lactation. If a high proportion of cases occur in later lactation, this suggests that either the environment could be a risk factor or it could be due to repeat cases from *Strep uberis* or *Staph aureus*.

Analysis of clinical mastitis by stage of lactation is only likely to be able to be extracted from herds which record on a computer system which can analyse this mastitis data easily.

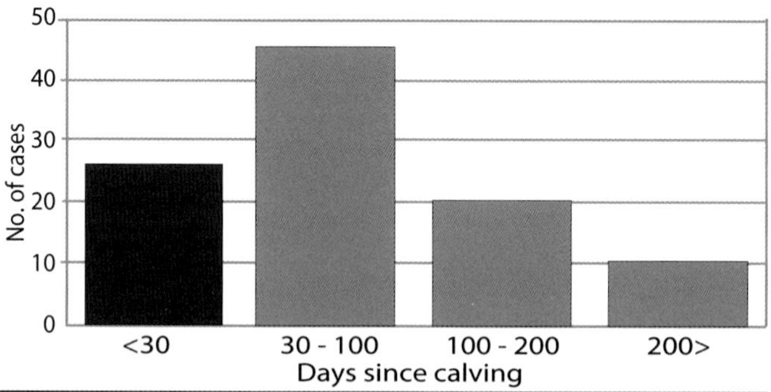

Herd 1: Percentage of mastitis cases by days in milk

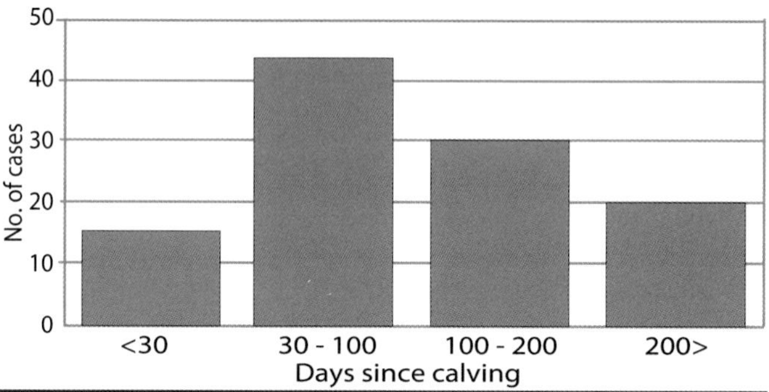

Herd 2: Percentage of mastitis cases by days in milk

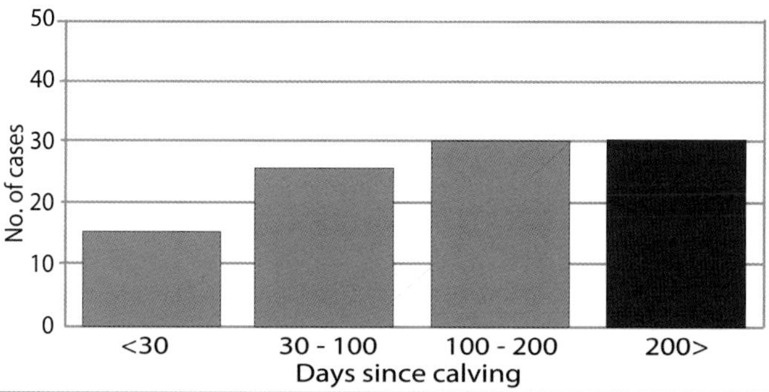

Herd 3: Percentage of mastitis cases by days in milk

FACT

- 50% of all clinical cases occur within the first 100 days of calving

Herd 1

Data from herd 1 shows that 25% of all clinical cases are occurring within 30 days of calving. This suggests problems with cows at or around calving and might also indicate problems with dry period infections.

70% of all cases are occurring within the first 100 days of lactation. This is higher than the expected 50% of cases. Dry period & early lactation infections are suspected as being a problem.

Herd 2

The chart shows that 50% of all mastitis cases are occurring within the first 100 days of lactation. Only 8% of cases are occurring within the first 30 days of calving. This is a typical example of what you would expect from a herd with good distribution of mastitis cases.

Herd 3

30% of all cases are occurring 200+ days in lactation which is far higher than expected. 60% of all cases are occurring more than 100 days in milk. The high proportion of cases 200+ days could be due to either problems with repeat cases of mastitis, as commonly can occur with *Staph aureus* and *Strep uberis*, or very poor environmental conditions in late lactation.

Mastitis parameters

Incidence by house, shed or group

Recording incidence by group or house can help highlight which, if any, has a mastitis problem. This can be of great value in robotic herds where environmental conditions, stage of lactation and stocking density may vary between individual groups.

One robotic herd has three groups on cubicles; a heifer group and a high and low yielding group. A fourth group is housed on a straw yard system. Individual records show that the highest incidence is in the straw yard cows and the lowest is in the heifer group.

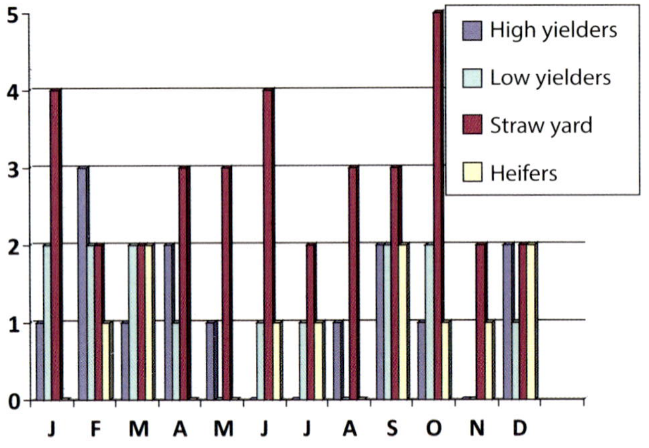

This graph clearly shows that the cows on the straw yard (red bars) have the highest incidence of mastitis while the heifer group (yellow) have the lowest incidence. The stocking density of the four robots is the same and so the differences are the environment and other factors such as milk flow rates with high yielding cows.

Sometimes, you do not even have to analyse the data as the farmer will be able to tell you that the majority of mastitis originate from one house. For example, the high yielding cows were housed in a cubicle shed which had mats and dried sawdust as bedding. Cows in later lactation were housed in a modified cubicle shed and bedded on deep sand with low levels of mastitis. Mastitis levels were high in cows on cubicles bedded with dried sawdust.

The groups were swapped over and the incidence of clinical mastitis in the high group dropped significantly. The number of cases in the later lactation cows increased slightly due to being put into a higher risk environment. This proved to be a highly successful measure to reduce clinical mastitis.

Clinical Mastitis Classification

(According to severity

Grade 1:
Clots or changes seen in milk

Grade 2:
Clots or changes seen in milk and a swollen quarter

Grade 3:
Sick cow with clots or changes seen in milk

It should be remembered that the majority of cases of clinical mastitis are mild.

Recording the severity of mastitis can be reassuring in confirming this. If a herd has a high incidence of grade 3 mastitis then this suggests problems with *E. coli* or *Strep uberis* mastitis.

Mastitis parameters

Milking cow tube usage

The number of milking cow tubes used will depend on the number of cases of mastitis, the number of tubes used to treat each case and whether any high cell count cows were treated with milking cow tubes. Most pharmaceutical companies recommend 3 tubes per case but the average usage in the UK is between 4 and 5. The target is under 1.5 milking cow tubes per cow per year.

A high number of milking cow tubes per cow per year suggests one or more of the following:

- High incidence of clinical mastitis
- Failure to record all clinical cases
- Poor response to treatment
- Treating high cell count cows
- Mastitis tubes used for other reasons

Most vet practices have computerised accounting systems that can give the number and type of intramammary tubes supplied over any period of time, so this information should be easy to obtain.

$$\text{No. milking cow tubes per case} = \frac{\text{No. of milking cow tubes used over a year}}{\text{No. of mastitis cases in the same period}}$$

Worked example

A herd of 150 cows had 30 recorded cases of clinical mastitis in the past year. 450 milking cow tubes were used in the last year.

$$\text{No. milking cow tubes per case} = \frac{\text{450 (milking cow tubes used)}}{\text{30 (cases of clinical mastitis)}} = \textbf{15 tubes /clinical case}$$

$$\text{No. tubes per cow per year} = \frac{\text{No. tubes used per year}}{\text{Average number of cows}}$$

$$\text{No. tubes per cow} = \frac{\text{450 tubes}}{\text{150 cows}} = \textbf{3 milking cow tubes/cows/year}$$

For this herd, 15 milking cow tubes per clinical case is abnormally high at over three times the average use and the total tubes used per cow per year is 3 which is twice the target figure. It is suspected that this herd does not record all clinical cases and that the incidence of clinical mastitis is much higher. There could be treatment issues as well.

FACT

- Between 4 and 5 milking cow tubes are commonly used for each case of mastitis

Benchmarking of clinical mastitis

Comparing mastitis incidence between herds

It is useful to compare mastitis incidence or milking cow tube usage between herds. A traffic light system can be used where green indicates levels below target; Amber is where targets exceeded and red where there is urgent need for action to be taken. The two bar charts below show the mastitis rate and milking cow tubes per cow per year for 20 herds.

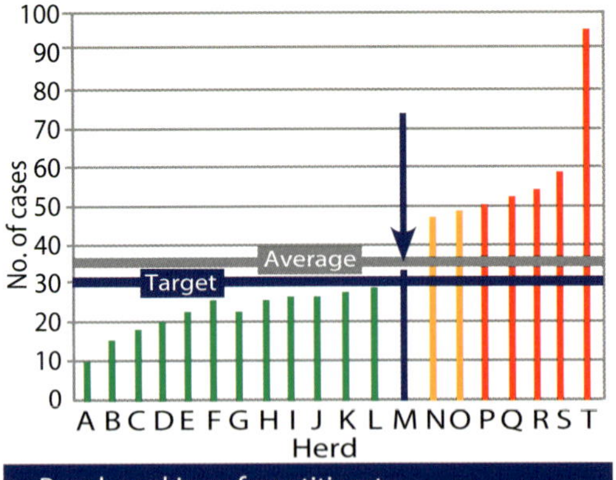

Benchmarking of mastitis rate

The first benchmarking graph shows the mastitis rate of twenty different herds (A to T). The mastitis rate varies from 10 up to 95 with an average of 38, shown as the grey 'average' line. Herd M is shown in dark blue with a mastitis rate of 31. The herds with green bars have a mastitis rate of under 30, orange up to 50 and red above that.

This chart shows that 60% of herds were below the target level. High incidence herds see others have a much lower incidence. This is often a good motivational tool for people to start making improvements.

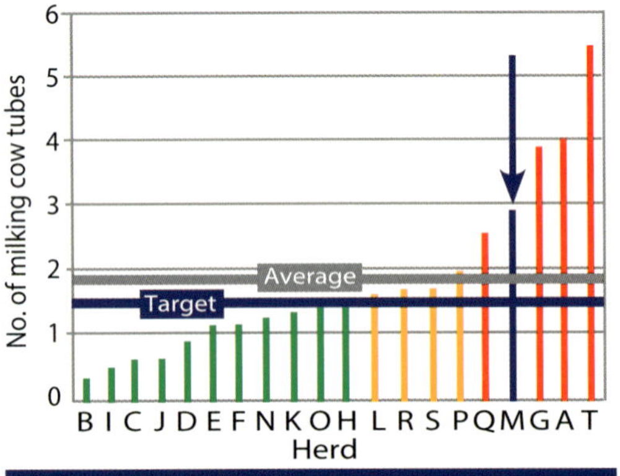

Milking cow tubes per cow per year

The second graph shows the number of milking cow tubes used per cow per year for the same twenty herds. This varies from 0.3 to 5.5 with the average of 1.8 shown in the grey 'average' line. The target figure is 1.5 show in the blue 'target' line on the left.

The order of these herds has changed as the number of tubes per cow per year does not necessarily relate to the mastitis rate. Herd M, which recorded a mastitis rate of 31 has the fourth high tube usage at 2.8 tubes per cow per year (dark blue bar and blue arrow). This indicates that an average of 9 tubes per clinical case were used (2.8 tubes x 100 cows = 280 tubes divided by 31 cases of clinical mastitis per 100 cows is just over 9 tubes per case.

This suggests that either not all cases of mastitis have been recorded, there are problems with treatment or some high cell count cows are being treated during lactation.

These two graphs are very useful when used together. The milking cow tubes per cow per year is a very useful benchmark and cannot be manipulated whereas it is easy to under-record clinical cases. It is possible to have a low mastitis rate by not recording cases. It is impossible to hide milking cow tube used per cow per year.

Benchmarking is a very good motivational tool for herds and farm staff which have successfully reduced clinical mastitis. Farms can see and enjoy their success. Likewise, herds which have a high incidence of clinical mastitis and consider this to be normal can see that others are achieving much lower rates. This encourages them to look at ways in which this can be achieved.

Herd Examples

Practical use of records

All three herds have 100 cows and a mastitis rate of 60 cases/100 cows/year; twice the target level of 30. The herds vary in their management and calving patterns. These examples show how data can be used to help identify the cause of clinical mastitis. Findings will always need to be confirmed by bacteriology and from further investigations.

	A	B	C	Target
Mastitis rate	60	60	60	<30
Herd cell count	125	250	250	<150
Percent of herd affected	45%	30%	18%	<20%
Recurrence rate	5%	18%	22%	10
Tubes/case	3	6	8	3-5
Seasonality of mastitis	Seasonal (Jul-Oct)	All year round	All year round	
Percent of cows with mastitis in first 30 days of lactation	14%	7%	5%	<8%

NB: Figures highlighted in red in the table above are above target levels

Healthy cows on clean pasture will have a low risk of clinical mastitis

Herd Examples

Herd A

Target 1: A dirty calving pen will increase mastitis

This is an autumn calving herd where cows are housed on cubicles in the winter. The herd has overstocked calving paddocks and pens at housing. The distribution of clinical mastitis is shown in the bar chart below and most mastitis occurs from July onwards when calving starts and through the housed period. The cell count is under 150 and this indicates that it's unlikely that there are significant *Staph aureus* or *Strep uberis* infections.

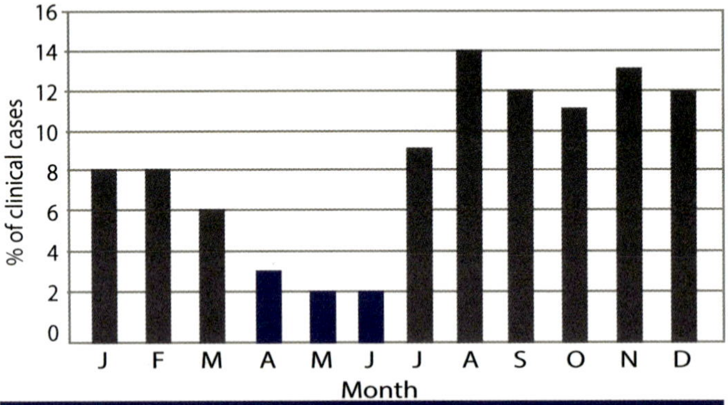

Herd A - Percent clinical cases by month

Contagious mastitis is highly unlikely to be contributing to clinical cases due to the low herd cell count. These data suggest that environmental mastitis is the most likely cause with infections originating from the dry period and lactational infections occurring when housed. *E. coli* is likely to be a problem as the low cell count suggests that *Strep uberis* infections are low.

- 45% of cows have had one or more cases of mastitis indicating environmental infections
- The recurrence rate is 5% indicating a good response to treatment
- An average of 3 tubes per clinical case which is low
- 14% of cows have had mastitis within the first 30 days of calving, almost twice the target figure and indicates dry period infections

Herd examples

Herd B

Target 2: This straw yard is very overstocked

Cows calve all year round and are housed in straw yards during the winter months. The farmer comments that he gets more mastitis when the cows are dirty at housing.

The distribution of cases is shown in the bar chart below and shows that most mastitis is during the housed period from October to April.

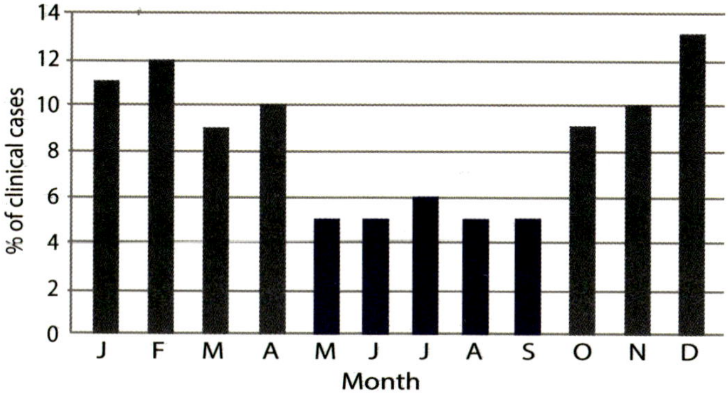

Herd B - Percent clinical cases by month

These data suggest environmental mastitis which is being picked up during lactation. This is most likely to be due to *Strep uberis* which is commonly associated with straw yards. However, there could also be some contagious bacteria such as *Staph aureus* causing clinical mastitis. Bacteriology will confirm the cause.

- The cell count is high at 250 and indicates that there could be *Staph aureus* or *Strep uberis* infections
- 30% of cows have had one or more cases of mastitis indicating environmental infections
- The recurrence rate is 18% indicating a poor response to treatment, which would be common with *Staph aureus* or *Strep uberis* mastitis
- An average of 6 tubes per clinical case which suggests treatment success is poor
- 7% of cows have had mastitis within the first 30 days of calving, which is under the target figure, and indicates no significant problems with dry period infections

Herd examples

Herd C

Target 3: Excellent housing conditions, wide passageways and clean cows

This herd calves all year round and has very good environmental conditions. The cell count is high at 250 indicating problems with contagious mastitis such as *Staph aureus* or *Strep uberis*.

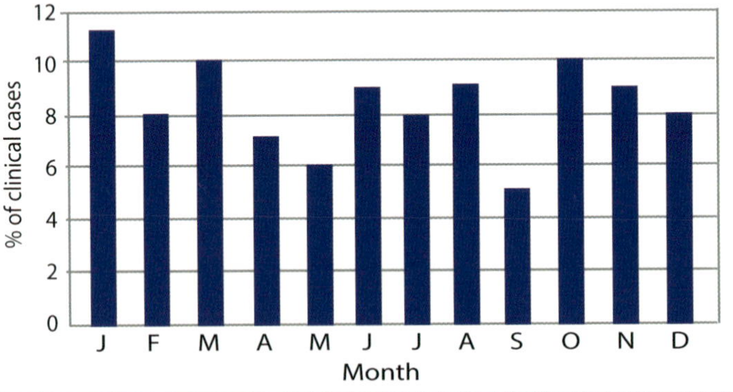

Herd C - Percent clinical cases by month

The high cell count suggests that there are problems with *Staph aureus* or *Strep uberis*. However, the percent of the herd affected and the fact that there is no seasonality of mastitis at housing suggests that *Staph aureus* is the more likely cause. This is supported by the high recurrence rate and the high number of tubes used per clinical case. These data indicate contagious mastitis infections entering during lactation.

- The cell count is high at 250 and indicates that there could be *Staph aureus* or *Strep uberis* infections
- 18% of cows have had one or more cases of mastitis indicating no significant problems with environmental infections
- The recurrence rate is high at 22% indicating a poor response to treatment, which would be common with *Staph aureus* or *Strep uberis* mastitis
- An average of 8 tubes per clinical case which suggests treatment success is very poor
- 5% of cows have had mastitis within the first 30 days of calving and indicates no significant problems with dry period infections

Summary of parameters and targets

Key targets

Parameter	Target	If over target suspect
Mastitis rate	<30	Clinical mastitis problem
Percent of herd affected	< 20%	Environmental mastitis
Recurrance rate	< 10%	***Strep uberis*** or ***Staph aureus*** mastitis Poor mastitis treatment Late detection of clinical cases
Mastitis cases within 30 days of calving	< 8%	Dry period infections
Milking cow tubes per cow per year	< 5	High levels of clinical mastitis Poor response to treatment Treating high cell count cows Milking cow tubes used for other things
Seasonal variation		Environmental conditions could be a problem

Table of targets and parameters

Bacteriology

How to control clinical mastitis

4

Introduction

Types of mastitis bacteria

It is important to know which bacteria are responsible for causing mastitis in your herd. This will help you fine tune your control measures and ensure that treatment success will be maximised. All mastitis bacteria are different.

The majority of clinical mastitis is due to environmental organisms such as *E. coli* and *Strep uberis*. With environmental organisms, the reservoir of infection is the environment itself. Environmental bacteria are transferred onto the teats between milkings whenever cows lie down in dirty conditions or from splashing of manure onto the teats. Infection can enter through the teat canal during milking, between milkings and during the dry period.

Infections can enter during milking if the teats have not been properly cleaned and dried before the unit is attached. Infections can occur between milkings if cows lie down immediately after milking or lie down on very dirty beds between milkings. Dry period infections can occur because not all teats fully seal throughout the dry period (see the 'Dry period infections' section).

FACTS

- The majority of clinical mastitis is due to environmental organisms

Staph aureus is another cause of clinical mastitis. This is a contagious mastitis bacteria. Contagious bacteria live in the udder and on the teats and are spread from cow to cow at milking time only. Infections establish on the teat and teat canal and then can penetrate the mammary gland.

Comparison of contagious and environmental infections

	Contagious	Environmental
Bacteria	*Staph aureus* *Strep uberis* *Strep agalactiae* CNS (Coagulase negative Staphylococci) *Corynebacterium bovis* (*C. bovis*) *Mycoplasma*	*E. Coli* *Strep uberis* *Klebsiella* *Pseudomonas* *Strep dysgalactiae* Yeasts and fungi *Prototheca*
Source of infection	Udder	Environment
When spread	During milking	Between milkings During milking During the dry period
Type of mastitis seen	Predominantly subclinical Clinical cases from *Staph aureus, Strep uberis and Mycoplasma*	Mainly clinical Subclinical for *Strep uberis*
Effect on herd cell count	Raised	None unless *Strep uberis* problems

Introduction

Mastitis bacteria in the udder

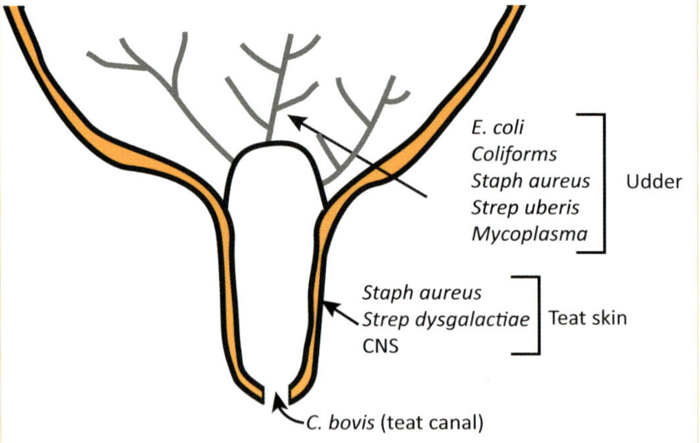

Location of bacteria in the udder

Where do the mastitis bacteria live in the udder?

The different bacteria live in different udder locations. *Staph aureus* lives deep in the udder tissue while *E. coli* and the coliforms live in the milk and milk ducts. This has great significance when it comes to treatment as it is much easier to kill bacteria floating around in the milk as they are more likely to come into contract with intramammary antibiotics. Bacteria which bury themselves deep in the udder tissue will be hidden away from white blood cells and antibiotics and so these infections are more likely to persist and be chronic.

	Milk and in ducts	Deep in the udder tissue
E. coli and the coliforms	+++	-
Strep uberis	+++	+
Strep aureus	+++	++
Strep dysgalactiae	+++	+
CNS (coagulase negative straphs)	+	+
Mycoplasma	+++	++

Coliform bacteria

E. coli

Coliforms are a group of environmental bacteria that include *E. coli, Klebsiella* and *Enterobacter. E. coli* is the most significant coliform bacteria that cause clinical mastitis in cows.

E. coli is the commonest cause of clinical mastitis throughout the world. This is because this pathogen is so prevalent throughout the environment and is shed large numbers in cow manure, so anywhere the teat comes into contact with manure there is a risk of *E. coli* mastitis. Cows which end up lying in dirty conditions will be more prone to *E. coli* infections.

People commonly associate *E. coli* mastitis with toxic cases and death; however the majority of cases are mild. The range of severity of *E. coli* mastitis is shown in the figure below. The top of the triangle shows the most severe cases where the cow is very ill and toxic. The majority of clinical cases are mild. About 80% of *E. coli* infections undergo self-cure and are eliminated by the cow's immune system without showing any clinical signs.

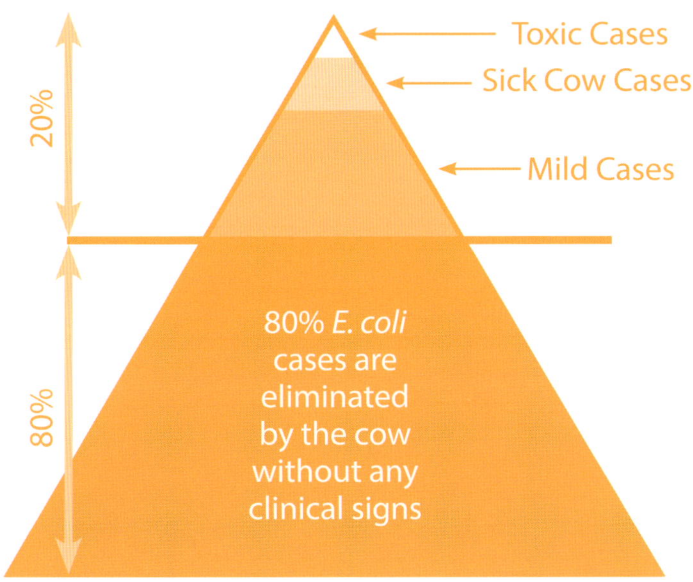

E. coli grows rapidly and can double every 20 minutes. Often, *E. coli* is not isolated from bacteriology samples by the time clinical mastitis is seen. This is because it is the toxins and not the bacteria which are responsible for the signs of clinical mastitis. *E. coli* die and release the toxin and it is this and not the bacteria which are responsible for clinical signs.

FACTS

- *E. coli* is the commonest cause of clinical mastitis throughout the world
- Produces endotoxins
- Clinical mastitis only
- Environmental bacteria
- Dry and lactational period infections
- Most cases are mild
- Up to 80% cases self-cure
- Can cause toxic mastitis and death
- *E. coli* can double every 20 minutes

Coliform bacteria

E. coli

E. coli clinical mastitis results in a very rapid rise in cell count. This is due to the speed at which they multiply. The cell count then drops back to normal after the bacteria have been eliminated. *E. coli* does not adhere to the udder tissue. This is a reason why you don't get chronic cows or recurrent clinical mastitis. Subclinical *E. coli* cases are uncommon.

Clinical cases can produce a watery secretion but you cannot rely on this as a way of diagnosing *E. coli* infections. You need to carry out bacteriology. Many samples come back with 'no growth'. This is to be expected as the bacteria have left and only the toxins remain.

FACTS

- Subclinical *E. coli* mastitis is uncommon
- The cell count with *E. coli* mastitis can be 10 million!
- Many bacteriology samples from cows with *E. coli* mastitis come back culture negative

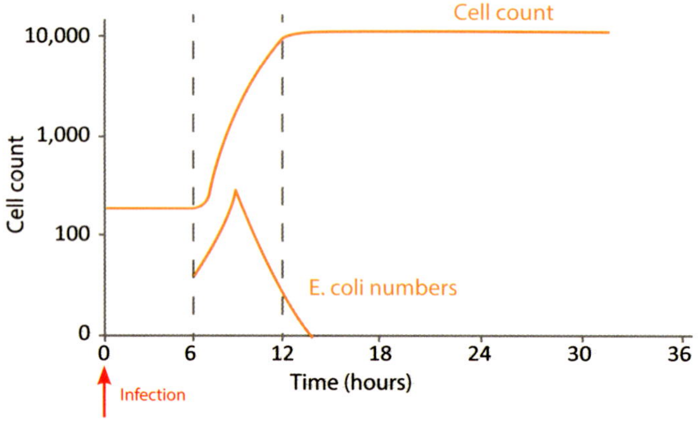

Good immune response in a mid lactation cow leads to a rapid elimination of *E. coli*

The diagram above shows that when a healthy cow is infected with *E. coli* the cell count can increase from 100 to over 10,000 within 12 hours. This rapid rise causes a reduction in *E. coli* numbers as the bacteria have been eliminated by the white blood cells.

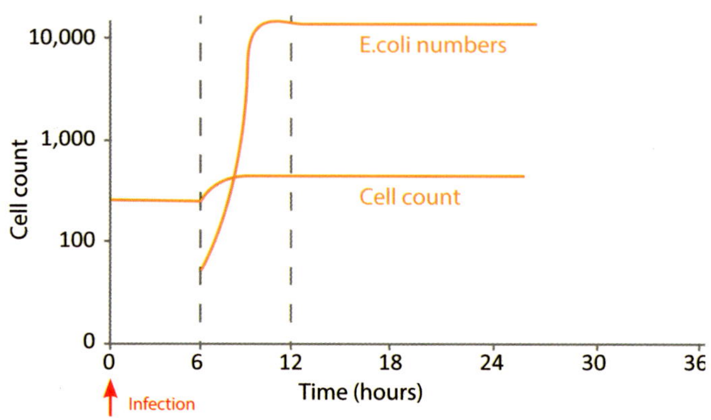

A very poor immune response allowing *E. coli* to multiply

The immune response in cows after calving can be very different. For unknown reasons, some freshly calved cows do not recognise *E. coli* invasion and there is no immune response. This allows the bacteria to multiply causing a severe toxic and often fatal mastitis. The cows that live often lose the affected quarters while others may remain ill for some time and eventually may be culled.

This effect is shown in the figure left, where there has been hardly any increase in somatic cells after infection. This allows *E. coli* to increase to levels of 100 million per ml of milk! Compare this to the normal response shown in the healthy cow.

Coliform bacteria

E. coli

This cow has just calved and has *E. coli* mastitis. She can't get up

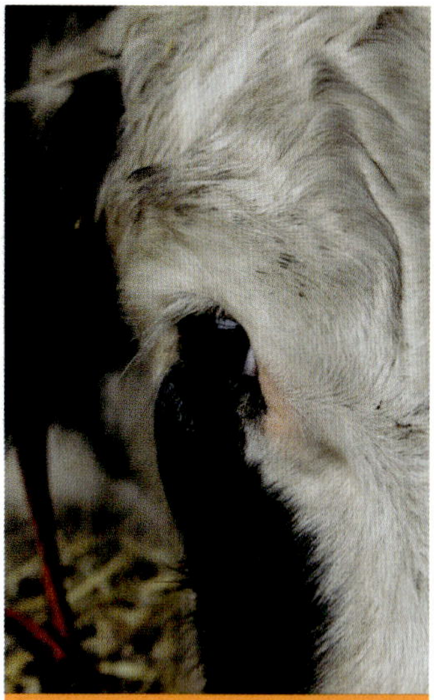

The eye is sunk into the head indicting dehydration which is common with toxic *E. coli* mastitis

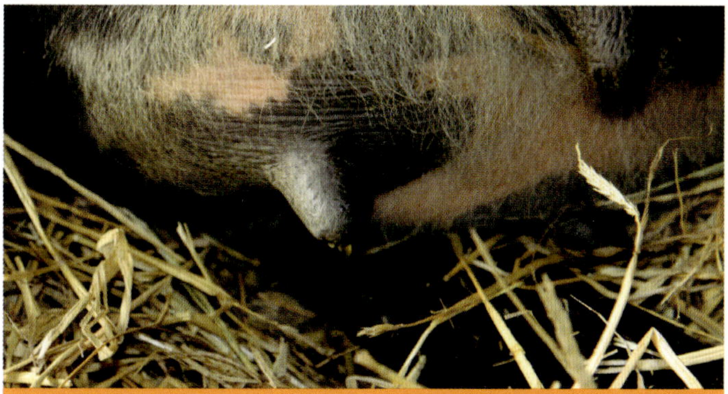

You can see the serous drop from the back right teat

This cow faces a massive *E. coli* mastitis challenge as her teats are buried in mud and manure

Many cases of *E. coli* produce a watery secretion

Coliform bacteria

Klebsiella

Klebsiella is commonly associated with bedding materials made from wood such as sawdust or wood shavings. *Klebsiella* is found everywhere in the environment, soil, water and even animal feed. Cows shed *Klebsiella* in their manure. 15% of humans excrete *Klebsiella* in their faeces. There are numerous strains and it is a normally found inside trees and is part of the decaying process of wood. However, *Klebsiella* thrives in sawdust, especially fresh sawdust which contains about 35% moisture.

FACTS

- *Klebsiella* is found everywhere in the environment
- In sawdust, levels of *Klebsiella* can be 100 times that of straw or sand bedding
- Ubiquitous
- Associated with wood bedding products
- Thrives in fresh sawdust
- Can cause toxic mastitis
- Outbreaks can occur in clean cows

Klebsiella is part of the tree decaying process

Klebsiella is shed in cow manure and common alleyways and walkways

Klebsiella are associated with trees

Klebsiella do not colonise teat skin and so are opportunists which get into the udder. The greater the exposure to the teat end, the higher the risk of mastitis occurring. The risk of environmental mastitis from *Klebsiella* is far higher in cows with dirty teats.

Coliform bacteria
Klebsiella

Research work shows that even if you predip, there is a much greater incidence in *Klebsiella* from cows that had dirty teats before milking than clean teats. You can see cows with three different cleanliness scores, 1 to 3 where 3 are the cows with the dirtiest teats. This shows that almost half of all teats are still contaminated with *Klebsiella* despite predipping before milking

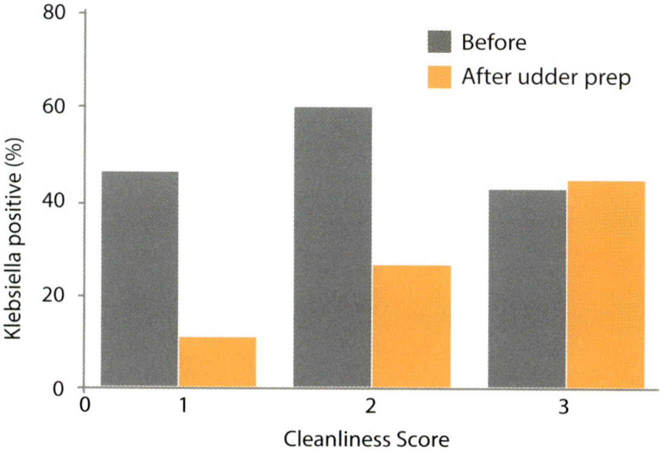

Effect of predip on cows with varying cleanliness scores

The graph below shows the levels of *Klebsiella* associated with different bedding types. This uses a log scale and so this means that the levels of *Klebsiella* in sawdust were almost 100 times that of straw, sand or crushed limestone bedding.

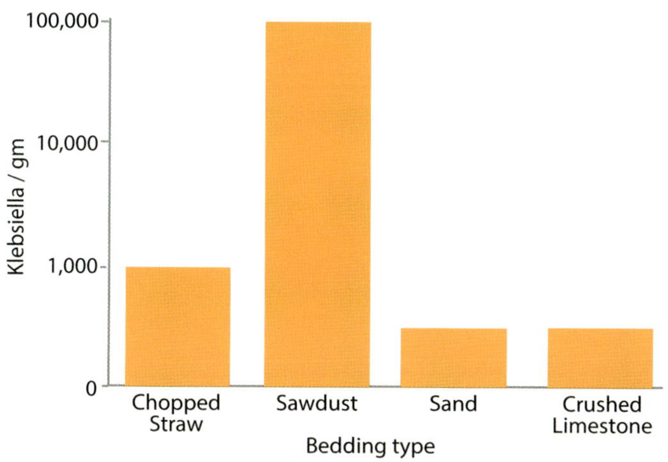

Klebsiella in bedding material

Outbreaks of *Klebsiella* are frequently associated with the use of sawdust as bedding material. The risk increases if fresh sawdust is used, or if dry sawdust becomes contaminated with water allowing fermentation with *Klebsiella*.

If you put your arm into a pile of fresh sawdust you will often feel the heat from fermentation. This will cause *Klebsiella* numbers to increase to very high levels. *Klebsiella* does not cause significant contamination to straw or sand bedding.

As sawdust becomes contaminated there is an increasing risk of *Klebsiella*

The majority of *Klebsiella* infections respond quite well to treatment, however some infections can persist and respond poorly. This is because different strains are present on most farms. In severe cases, *Klebsiella* can cause toxic mastitis resulting in death.

Strep uberis

Introduction

Strep uberis lives in the udder, mouth, between legs and other parts of body. It is also found in contaminated pasture, such as paddocks and walk ways, and grows well in straw bedding. *Strep uberis* was isolated in 89% soil samples, 23% manure samples and 63% environmental samples. It can be introduced onto pasture by cows and will disappear about 3 weeks after cows leave the pasture.

Cows introduce *Strep uberis* onto pasture

Strep uberis is a range of bacteria with some strains being more resistant to being engulfed by white blood cells. These strains can persist making them difficult to cure. Other strains can survive inside cells where they are protected from the action of antibiotics. *Strep uberis* is easy to diagnose as it is shed in high numbers. These bacteria live in the milk and in the milk ducts.

Strep uberis causes clinical and subclinical mastitis. Clinical mastitis can originate from subclinical infections and vice versa. It is unusual in that it has both contagious and environmental properties. It is commonly associated with housing systems that use straw as bedding, especially in straw yards systems or loose housing. Outbreaks of clinical mastitis can also occur at pasture. Infections can enter the udder during the dry period.

Straw that becomes contaminated with cow manure will ferment and this will provide everything that bacteria need to grow; heat, moisture and nutrients

FACTS

- Clinical and subclinical infections
- Contagious and environmental properties
- Outbreaks at pasture
- Associated with straw bedding
- Dry period infections
- Repeat clinical cases
- Can persist from lactation to lactation

Some *Strep uberis* strains are intracellular and protected from antibiotics. The glass jar is the cell wall and the sweets inside are the bacteria

Strep uberis

A common cause of clinical mastitis

Strep uberis is a common cause of clinical mastitis and some cases result in high temperatures. Repeat cases of clinical mastitis are common, just as they are with *Staph aureus* mastitis. Clinical *Strep uberis* infections can become chronic resulting in high cell counts. Cows that have subclinical infections can also develop clinical mastitis. *Strep uberis* can persist from one lactation to the next and the response to dry cow therapy is variable.

Prevalence of clinical mastitis will be related to the presence of risk factors. Cows often shelter under trees in wet or hot summers. This can cause a build-up of *Strep uberis* numbers on soil on which cows are lying and can cause clinical cases.

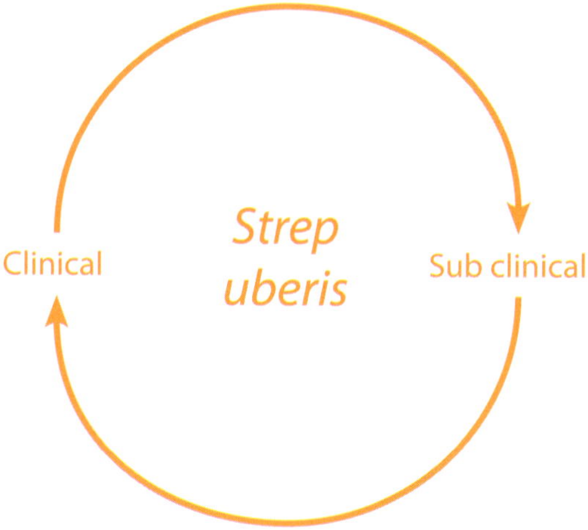

FACTS

- *Strep uberis* was isolated in 89% soil samples and 23% manure samples
- Found on pastures 3 weeks after cows started grazing
- *Strep uberis* is a range of bacteria
- *Strep uberis* clinical mastitis can originate from subclinical infections and vice versa
- Can become chronic and result in high cell counts
- Expect *Strep uberis* mastitis from cows in straw yards
- There can be 1 million *Strep uberis* in a gram of dirty bedding

These cows are sheltering from the sun on a very hot summers day

Cows sheltering from rain under the same trees where these is a risk of *Strep uberis* when they lie down

Strep uberis

In the environment

This shows dirty conditions when cows were sheltering under this tree

If cows are kept in dirty straw yards then outbreaks of clinical mastitis are likely to occur. You need to use plenty of clean dry straw and bed up daily to try and minimise *Strep uberis* mastitis in straw yard systems. There can be over one million *Strep uberis* bacteria per gram of contaminated straw bedding!

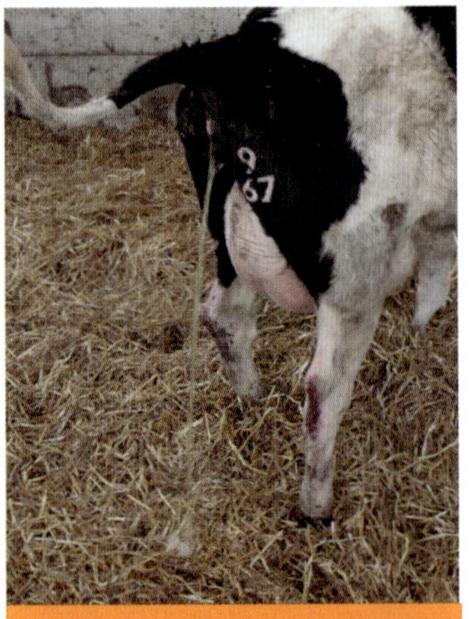

Cows are very wet animals and will produce an average of 50 litres of moisture each day in the form of faeces, urine and respiration. When you walk around a poorly drained straw yard you can often find that it sounds squelchy under foot

The risk of *Strep uberis* mastitis can be very high in a straw yard as cows can lie anywhere and much of the yard will become contaminated

This straw yard is very overcrowded and the risk of *Strep uberis* mastitis will be high

Strep uberis

Dry period infections

Dry period infections tend to enter the udder in the first week after drying off and the two-week period before calving. These often do not cause problems at the time. Dry period infections cause clinical mastitis in the first 100 days of lactation. Dry period *Strep uberis* infections can occur even if antibiotic dry cow therapy has been administered.

Cure rates for *Strep uberis* infections can be variable. Clinical mastitis cure rates using antibiotic intramammary tubes are commonly about 60 – 70%. Prompt treatment is necessary to overcome those strains of *Strep uberis* that are not engulfed by white blood cells.

FACT

- *Strep uberis* can be a common cause of repeat cases of clinical mastitis
- *Strep uberis* can be spread to the next 3 cows on liners

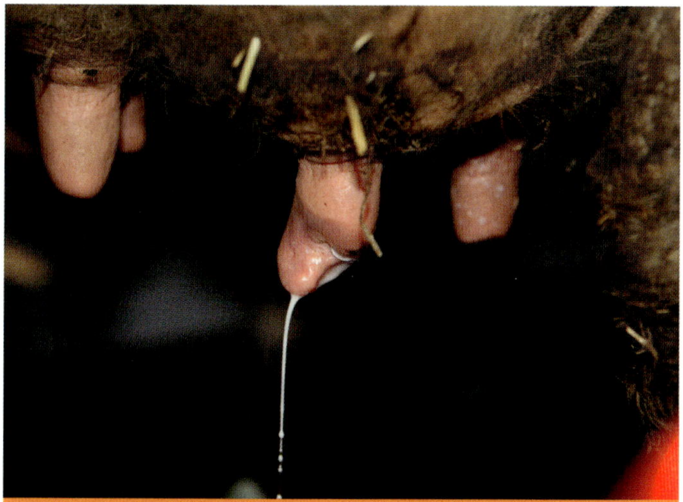

Cows that leak milk will have a far higher risk of picking up *Strep uberis* and coliform mastitis infections

Staph aureus

Introduction

Staph aureus causes clinical and subclinical infection. It is a very common cause of subclinical infection throughout the world. *Staph aureus* is a contagious bacteria that can be spread to the next 6 to 8 cows from a contaminated liner acting as a vector. It is spread in the milking parlour. *Staph aureus* is not spread onto teats from the environment, soil, air or water.

Staph aureus lives in the udder and on the skin of the teats. It thrives if teat skin condition is poor or damaged and easily colonises through wounds. *Staph aureus* and some other contagious mastitis bacteria like *Strep dysgalactiae* will all thrive on damaged teats and poor teat skin condition.

When *Staph aureus* enters the udder it buries itself deep in the mammary tissue. It is an intracellular bacteria as it can hide in cells. It commonly forms micro-abscesses and creates scar tissue as a result of tissue damage. *Staph aureus* results in chronic infections.

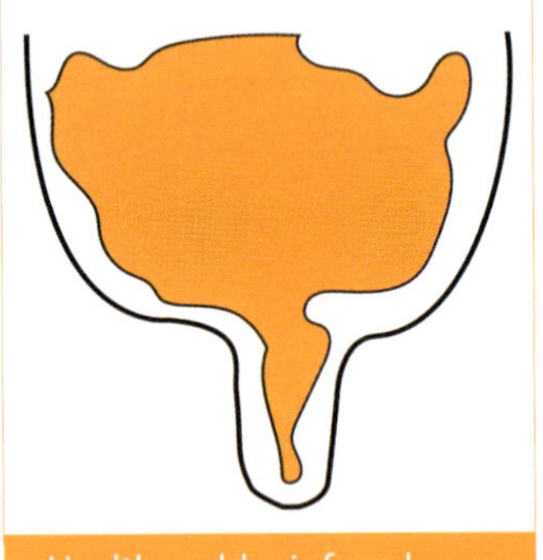

Healthy udder infused with dye showing excellent distribution

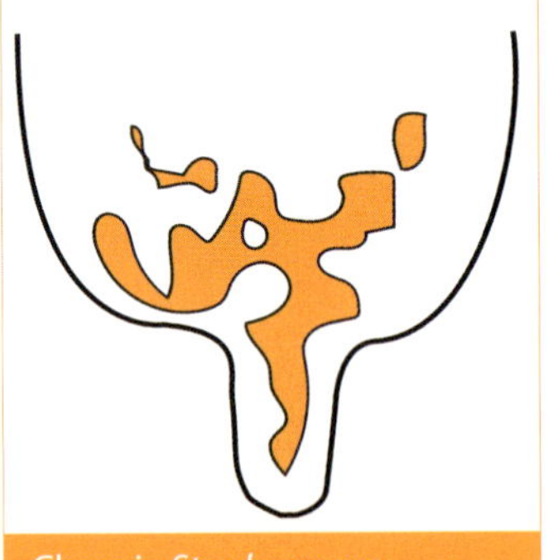

Chronic *Staph aureus* udder. Poor distribution so poor cure rates from tubes

The diagram above shows how well food dye is distributed throughout the udder after intramammary infusion in two animals. The udder on the left is from a healthy heifer and shows excellent distribution. This would be expected in an animal that has no subclinical or clinical mastitis history. In this case treatment can be successful.

The udder on the right is from an old cow with chronic *Staph aureus* infection. This shows very poor distribution which means that treatment success will be significantly reduced as antibiotic does not come into contact with the whole udder. Milk yield will also be reduced as if the dye cannot diffuse upwards due to damaged or blocked ducts then milk will not be able to be extracted from the udder.

FACTS

- *Staph aureus* can be spread to the next 6 to 8 cows on a liner
- It lives in the udder and on the skin of teats
- Contagious bacteria
- Clinical and subclinical mastitis
- Repeat cases of clinical mastitis
- Poor response to treatment
- Chronic infections
- Penicillin resistance strains
- Intermittent shedding
- Micro-abscesses in the udder
- Significant reduction in milk yield
- Culling of problem cows

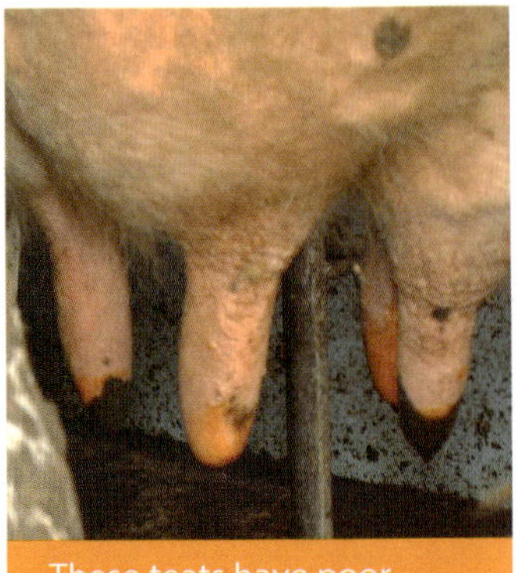

These teats have poor teat skin condition as the milker only coats the bottom of the teat with teat dip

Staph aureus

Micro-abscesses and acute Staph aureus

Micro-abscesses can be compared to a can of baked beans where the stainless steel can represents the wall of the abscess and the baked beans are the *Staph aureus* bacteria. These micro-abscesses break open releasing infection which can then be spread to other cows or different quarters in the same cow. This is equivalent to opening a can of baked beans.

Antibiotics are unable to penetrate the stainless steel wall of the micro-abscess or the cells where *Staph aureus* can hide. Beta lactamase producing strains of *Staph aureus* are resistant to penicillin. These are the Beta lactamase producing strains. 40% of UK *Staph aureus* strains are Beta lactamase producers. These are some of the reasons why treatment success with *Staph aureus* is so poor.

Staph aureus forms micro-abscesses and can be compared to a can of baked beans. The stainless steel can is the equivalent of the wall of the micro-abscess and cannot be penetrated by antibiotics. The baked beans are the *Staph aureus* bacteria

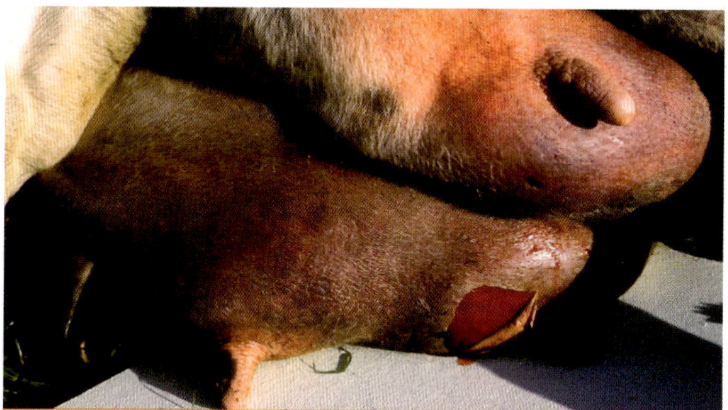

This cow had acute *Staph aureus* gangrenous mastitis in all four quarters and the toxaemia killed the cow within 24 hours

Staph aureus can cause acute mastitis where the toxins cause gangrenous mastitis which will result in death of the quarter and in severe cases can also kill the cow. Some cows with gangrenous mastitis can remain remarkably healthy and just lose the quarters.

Staph aureus causes significant damage to the milk producing tissue resulting in reduced milk production. Milk yield can be reduced by as much as 25% in chronically infected cows. Cure rates following antibiotic treatment are generally low with clinical mastitis cure rates averaging 25%. However, cure rates drop even more in cows with chronic infections.

There is no alternative except to cull chronic *Staph aureus* cows with either clinical or subclinical mastitis. These cows are unproductive and act as a reservoir of infection to the remainder of the herd. Prevention of *Staph aureus* infections is essential as treatment is generally unsuccessful.

FACTS

- *Staph aureus* is spread by liners, hands and common cloths
- You have to cull old chronic *Staph aureus* cows
- You cannot treat your way your of a *Staph aureus* problem
- *Staph aureus* can cause gangrenous mastitis
- *Staph aureus* can reduce milk yield by 25%

TOP TIP

- Cull old chronic *Staph aureus* cows

Strep dysgalactiae

Overview

Strep dysgalactiae is commonly called *Strep dys*. It is unusual in that it has both contagious and environmental properties with the most prominent property being environmental. It can cause both clinical and subclinical mastitis. *Strep dysgalactiae* is considered to be a minor mastitis pathogen.

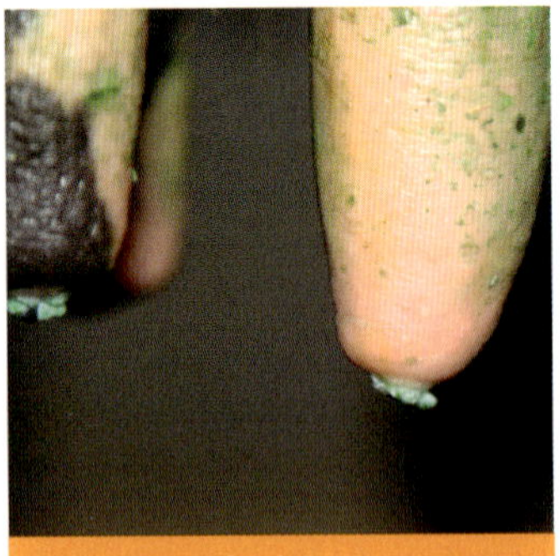

Expect to find *Strep dysgalactiae* with hyperkeratosis shown in green

It is most commonly associated with teat skin damage or hyperkeratosis, where the teat canal is pulled through the bottom of the teat. Many bacteria will thrive on damaged teat skin especially *Staph aureus* and *Strep dysgalactiae*.

Hyperkeratosis can be caused from a variety of reasons, most of which are induced by the milking machine. Common problems include over milking, too high a vacuum level and pulsation problems.

Response to treatment during lactation for clinical cases and antibiotic dry cow therapy for subclinical mastitis is good. It is controlled by the standard control measures for both contagious and environmental mastitis.

FACTS

- Contagious and environmental
- Clinical and subclinical mastitis
- Associated with teat damage
- Good response to treatment
- Minor mastitis bacteria
- Uncommon cause of high herd cell count

Coagulase negative Staphylococci

Overview

Coagulase Negative Staphylococci are commonly called CNS. They are all the *Staphs* apart from *Staph aureus*. There are about 50 different strains with new species still being found. Clinical mastitis due to CNS is not uncommon.

Approximately 10 species of CNS are known to cause mastitis. These strains can be difficult to differentiate and new technology is being developed to accurately identify individual strains. CNS are commonly found on the skin of teats and on our hands. These are perfectly normal findings.

Some strains of CNS cause clinical mastitis while others have a preventative effect. You find a greater proportion of CNS bacteria in herds with low cell counts. This is why it is called a minor mastitis pathogen. They are easily killed using antibiotic dry cow therapy.

FACTS

- CNS are a range of bacteria
- Minor mastitis pathogens
- CNS can help prevent mastitis
- CNS will not cause a high herd cell count

Pseudomonas aeruginosa

Environmental bacteria associated with water supplies

Pseudomonas aeruginosa is most commonly associated with contaminated water supplies. It is an environmental bacteria that can also be found in wet bedding. Outbreaks of *Pseudomonas* mastitis have occurred when people wash and do not dry teats before the unit is attached. This can allow bacteria easier access into the udder.

Contamination of water with *Pseudomonas* bacteria can occur on farms which use their own water supply, contamination in header tanks which store water for washing teats, or even water heaters. *Pseudomonas* can be present in parlour drop hoses even when disinfectant solutions are added to the wash water.

Predipping, where the teat is coated in a disinfectant solution for about 30 seconds which is then wiped off, will be an effective way to reduce the risk of *Pseudomonas* infections.

Pseudomonas mastitis has occurred from high levels of environmental contamination in wet bedding, wet areas in the environment, soil, overcrowding of cows, not removing contaminated bedding and allowing cows access to muddy areas or standing in water.

Outbreaks of *Pseudomonas* mastitis have also occurred following use of contaminated medicated towels used for disinfecting teats before infusion of intramammary syringes, or putting tubes in buckets of warm water before infusing. Some farmers do this to make it easier to infuse the tubes when temperatures are low. This organism can be difficult to remove and can survive in some disinfectant solutions.

Whenever *Pseudomonas* has been identified as a cause of clinical mastitis, all water sources should be tested immediately. You control *Pseudomonas* infections by reducing exposure to contamination. It is essential to ensure that all water used during milking is free from contamination.

Clinical signs can be variable from mild cases that respond well to treatment to acute cases resulting in death. Cows can also end up with chronic infections that respond very poorly to antibiotic therapy and many of these animals will be culled. This is because *Pseudomonas* bacteria can live inside cells where they are not accessible to antibiotics. It is resistant to many antibiotics.

FACTS

- Environmental bacteria associated with contaminated water supplies
- Can cause acute clinical cases and deaths
- Response to treatment is poor
- Chronically infected cows may need to be culled

TOP TIP

- Sample all water sources if you are getting *Pseudomonas* mastitis

Water hoses can be contaminated with *Pseudomonas*

Water header tanks and heaters are a common source of *Pseudomonas* infections

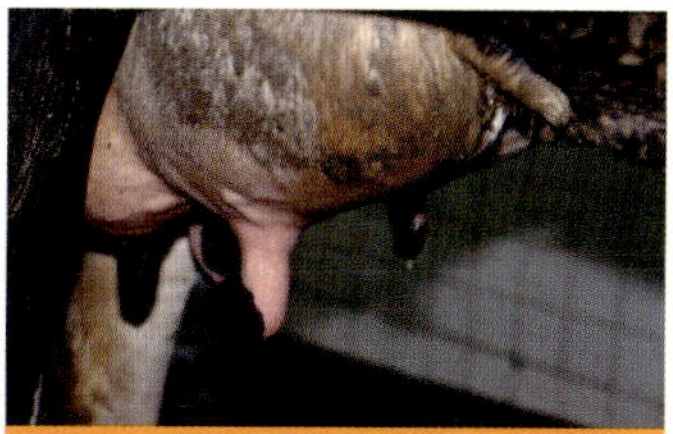

Pseudomonas outbreaks have occurred when people wash but don't dry teats

This herd had *Pseudomonas* mastitis following making up their own disinfectant paper wipes which were contaminated

Yeasts, fungi and moulds

Introduction

Yeasts and moulds are common environmental organisms; *Candida* is a yeast and *Aspergillus* is an example of a fungus. Infections will most commonly occur from poor intramammary infusion without properly disinfecting the teat end. These organisms are physically introduced through the teat canal on the end of an intramammary syringe.

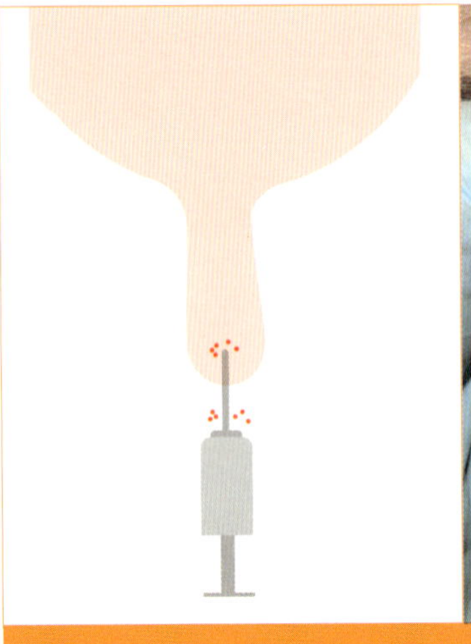
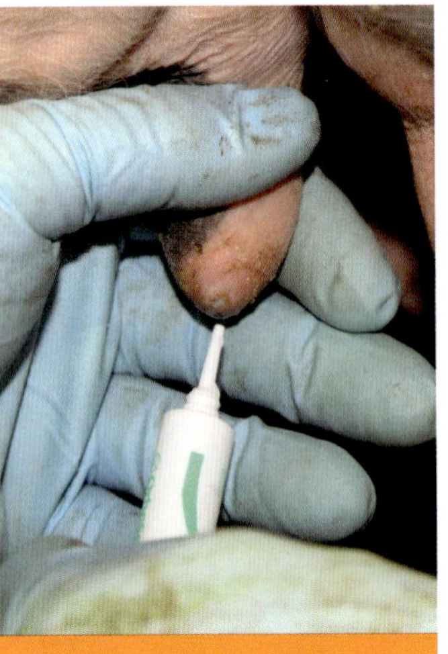

Yeast infections can occur from infusing tubes without disinfecting teats

Outbreaks of yeast mastitis have occurred following 'blitz therapy' where a group of cows are treated with intramammary antibiotics during lactation to eliminate *Strep agalactiae* infections. There can be a significant delay from when the organism is introduced into the udder until clinical signs are seen. Infected quarters may shed the organism for many months after returning to normal.

The risk of infection can increase if cows are allowed to lie on mouldy silage if this has been allowed to build up over a period of time. This risk rises further where there is poor pre-milking teat preparation. Cases have also occurred in herds which make home-made intramammary antibiotic preparations, as happens occasionally in developing countries where there may be no commercial intramammary antibiotic preparations available.

Yeasts and fungi tend to cause hard and swollen quarters that are very inflamed and painful. Animals can have a high temperature. Response to antibiotic treatment is ineffective as these are not bacterial infections. Some people have unknowingly treated clinical cases with antibiotics for a significant amount of time with no success whatsoever or even a deterioration in mastitis. This is because antibiotics eliminate any bacteria present in the udder allowing the yeasts, fungi and moulds to proliferate.

FACTS

- Yeasts are physically introduced into the udder on the end of an intramammary syringe
- Most commonly injected into the teat on a contaminated intramammary tube
- Do not respond to antibiotics
- Hard swollen quarters are common
- Unresponsive to treatment
- Chronic cases may be culled

TOP TIP

- Always disinfect teat ends before tubing any cow

Mouldy bedding like this could contaminate teats and cause yeast mastitis

Yeasts, fungi and moulds

Blitz therapy

It is possible to start off with a mild clinical case due to one of the common mastitis pathogens, and then introduce a yeast infection on an intramammary tube which is far more severe than the original infection. If yeast infections become chronic then the long-term prospects for the cow are going to be poor. Many end up being culled from the herd.

Infected animals should be milked last to reduce the risk of spreading infection from cow to cow. Unresponsive cases of clinical mastitis or cases with hard swollen quarters should be sampled to check if these organisms are responsible for the infection.

KEY POINT

- Sample cases of mastitis that are not responding to treatment

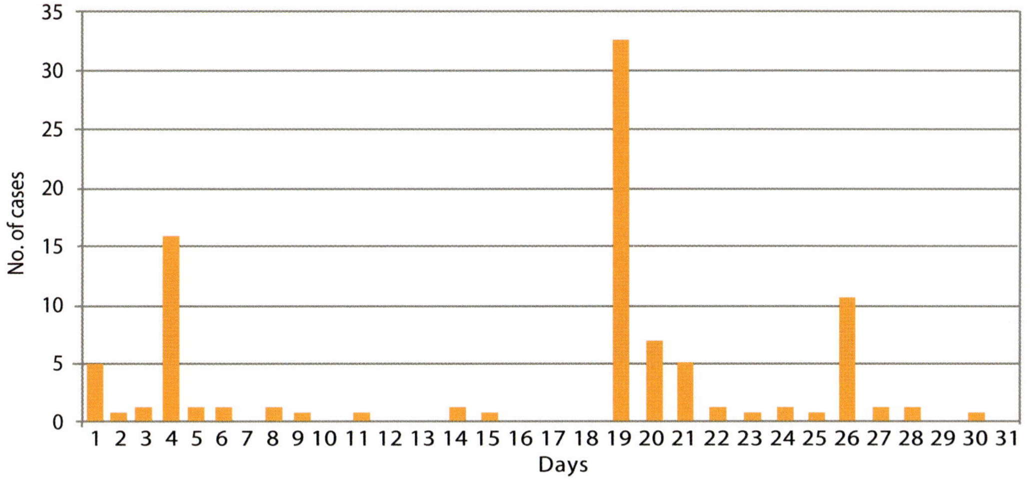

Number of clinical mastitis cases by days following blitz therapy

The chart above shows the number of cases of mastitis following blitz therapy which took place on day 0 in a herd of 200 cows. This shows an outbreak of clinical cases within a week of whole herd blitz therapy followed by a second wave of clinical cases 18 days after the blitz. All samples tested positive for the yeast *Candida krusei*. In this herd the teats were not disinfected before intramammary therapy. The farmer presumed that the end of the teats were clean after unit removal. Teat preparation was poor before milking and all teat ends were dirty. Yeasts were introduced into the udder on the intramammary syringe.

There was a very wide variation in clinical signs. Some animals were very sick and had four hard quarters; some quarters were not swollen and only had mild clinical signs.

Samples were collected from a variety of environmental sources including sawdust, other bedding materials and feedstuffs. All tested positive for the yeast called *Candida krusei* which was the same organism isolated from all the cases of mastitis.

Mycoplasma

Prevalence of Mycoplasma

Mycoplasma is a contagious form of mastitis that is increasing in incidence. The first recorded outbreak causing mastitis occurred in 1962 in a 50 cow herd in the United States. 25% of cows had mastitis that came back as culture negative using conventional bacteriology. Extended bacteriology testing identified *Mycoplasma bovis.*

Mycoplasma bovis is the most common strain. The bacteria are spread from cow to cow at milking time. As herds are expanding and buying in new animals, the risk increases even further. The percent of herds that tested positive for Mycoplasma is clearly associated with herd size as shown below. Over 20% of bulk tanks in US dairies with more than 500 cows tested positive compared to 2% in herds of 100 cows or less. Cell counts in infected animals are often over I million.

Prevalence varies greatly between countries and part of this will be whether laboratories look for the organism.

Country	Herd prevalence
US	about 8%
Europe	1-5%
Mexico, Iraq, Australia	55-100%
New Zealand	0%

Mycoplasma will spread from cow to cow during milking but also quarter to quarter through the lymphatic and blood systems as quickly as 4 days after the initial infection. Cows can become infected from calves and vice versa. Aerosol spread of infection occurs. The feeding of infected waste milk can also speed up spread of infection and milk from infected cows should not be fed to calves.

Mycoplasma look like fried eggs

FACTS

- Mycoplasma mastitis was first reported in 1962
- Over 20% of bulk tanks in US herds with more than 500 cows have tested positive for Mycoplasma
- Contagious mastitis
- Poor response to treatment
- Spontaneous cures can occur
- Cow to cow spread at milking
- Spread quarter to quarter through lymphatic system
- Screen bulk milk samples regularly

Mycoplasma

Testing and managing infections

Cows with Mycoplasma mastitis must be milked last. They must remain in the hospital pen until they are either proven to have cured or go to be culled. It does not really matter how much infection is spread within the hospital group as they will be separate from the main herd.

The importance of having a hospital pen cannot be over emphasised. Cows with Mycoplasma mastitis will spread infection to the rest of the herd if they are not kept isolated and milked last.

Treatment success is very poor and many cows end up being culled as they go off their food and dry up. It is advisable to test for this bacteria if cows are not responding to treatment. Most cows with Mycoplasma mastitis are perfectly healthy.

Mycoplasma is a difficult bacteria to grow as it is an anaerobic bacteria which means that it will only grow in the presence of carbon dioxide and takes about 7 to 10 days to grow. The true incidence of this organism is probably underestimated as it will not be identified using conventional bacteriology. There is a PCR test for Mycoplasma which will speed up diagnosis.

Mycoplasma is found in the respiratory and reproductive systems as well as in joints. Some cows with pneumonia or arthritis have developed clinical mastitis. Mycoplasma can be present in a herd without any clinical mastitis being seen.

The easiest way to screen a herd for Mycoplasma is through testing of bulk milk. You can pick up one infected cow in a herd of 300 or more from a bulk sample. This should be carried out on a regular basis as if these bacteria are present then the management of clinical mastitis cases needs to be changed. Herds must know whether they are free from Mycoplasma infection.

If a tank tests positive, another sample should be taken for confirmation. Infected cows can be intermittent shedders and so it can be difficult to detect all infected cows. It is best to test fresh rather than stored or frozen milk as this gives the highest recovery rates.

A good milking routine is essential to try and minimise any spread of infection. Gloves, post milking teat disinfection, individual towels or cloths and the use of steam back flush systems can prove very useful.

About 80% of infected cows develop clinical mastitis. Spontaneous cures do occur and these can only be confirmed following three negative samples. The speed at which spontaneous cures take place is variable and can take from one to more than three months.

Some herds use a test and slaughter policy where all cows are screened and those which test positive are culled. This is not to be advised in the case of outbreaks. Some herds might decide that they want to follow an eradication type policy, but many herds have been able to manage Mycoplasma through good management practices and only culling cows with persistent Mycoplasma infection.

FACTS

- You must ask your lab to test for Mycoplasma
- It takes between 7 and 10 days to grow on agar plates
- Always test fresh and not frozen milk

TOP TIPS

- Milk Mycoplasma cows last in the hospital pen
- Put mastitis cows in a hospital pen
- Regularly screen bulk milk
- Isolate positive cows

WATCH OUT!

- Double check positive bulk tank samples
- You do not have to follow a test and slaughter policy

Bacillus

Types of Bactillus associated with mastitis

There are two types of Bacillus that can cause mastitis; *Bacillus cereus* and *Bacillus subtilis*. Bacillus are found in the environment such as soil, water, vegetation and can also be found in manure, abscesses and wounds on the animal.

Bacillus cereus can cause gangrenous mastitis resulting in death of the animal. Outbreaks of Bacillus mastitis have been associated with cows lying on mouldy maize silage. Infections can also be introduced into the udder on the end of intramammary syringes if the teat ends have not been thoroughly disinfected before infusion. If people make home-made intramammary preparations, as can happen in developing countries, this can also be the cause of Bacillus mastitis.

FACTS

- Environmental bacteria
- Gangrenous mastitis

WATCH OUT!

- Disinfect teat ends before tubing cows
- Remove waste silage so cows cannot lie on it

Proteus

Overview

Proteus is commonly associated with bedding, feed and water. It is an uncommon mastitis pathogen but you can get herd outbreaks which are often chronic with severe clinical cases. There is a poor response to antibiotic therapy.

Little is known about how these bacteria cause mastitis but it is thought that they main risk is between milkings when the teat ends can become contaminated.

As Proteus is environmental in origin the common steps to avoid environmental mastitis seem to work well. Herds that have had cases should avoid overcrowding, keep cows as clean as possible, predip cows and use a quality post milking teat disinfectant.

FACTS

- Environmental organism
- Source is bedding, feed and water
- A contaminant

Prototheca

Prototheca Isolation Media

Prototheca was always considered to be an opportunist or exotic mastitis organism. Improved new diagnostic techniques have shown that its prevalence is significantly higher than originally thought. It is becoming increasingly common throughout the world.

Prototheca is an algae and can be associated with contaminated water supplies. There is no known effective treatment. It is commonly found in the environment and thrives in recycled sand. It has contagious and environmental properties and can be spread from cow to cow via liners. It is also thought to be spread through the lymphatic system from quarter to quarter. It thrives both in the environment and in the cows udder.

In 2010, mastitis testing labs started using Prototheca Isolation Media (PIM) and this increased recovery rates by a factor of 20. This indicates that this organism was significantly under diagnosed before then. It can be picked up through bulk tank screening. You only need one positive cow for the bulk tank to test positive in most herds. If a herd tests positive then all water sources on the dairy should be tested.

There can be a long lag time from when infection enters the udder to when clinical signs are seen. Unlike most other mastitis organisms, cell counts of infected animals can be as low, under 300. 80% of infected animals do not show signs of clinical mastitis.

There are different types of Prototheca strains but this is of no relevance at all from a control point of view. It is essential to identify infected animals and cull them from the herd. However, one of the difficulties is that these can be intermittent shedders and at times some infected cows are shedding less than 100 colonies/ml which can make diagnosis difficult.

Prototheca is antibiotic and heat resistant but can be killed with disinfectants such as teat dips. Likewise, back flush systems can help to reduce spread between cows.

FACTS

- Prototheca will be under diagnosed unless specialist media are used
- Found in the environment and thrives in recycled sand
- 80% of infected animals do not develop clinical mastitis
- Infected cows can have low cell counts
- Algae
- No effective treatment
- Subclinical and clinical infections
- Chronic infections
- Cow to cow spread during milking
- Cull infected cows

TOP TIPS

- Screen bulk milk for Prototheca
- Milk all infected cows last
- Cull infected animals

WATCH OUT!

- Some cows are intermittent shedders

Uncommon causes of mastitis

Quick reference list

Arcanobacterium pyogenes

A cause of mastitis in the dry period commonly called summer mastitis. It causes an acute and purulent mastitis where there is a poor prognosis once the infection is established. Flies are important in transfer.

Corynebacterium bovis

Can cause subclinical mastitis and raised cell counts. Has been associated with poor or delayed post milking teat disinfection. May also be isolated from the teat canal and not associated with mastitis. This bacteria is killed using dry cow therapy.

Leptospira hardjo

Seen in conjunction with abortions and milk drop as part of the leptospirosis complex. Very difficult to culture.

Listeria monocytogenes

Has possible human health implications and has been associated with soft or unpasteurised cheese.

Nocardia asteroides

Cows develop very hard quarters, high temperatures and there is extensive fibrosis of the udder. Environmental bacteria commonly associated with contamination when administering an intramammary tube. There is cow to cow spread as well as from the environment to cows. Poor response to antibiotics.

Pasteurella or Mannheimia species

Infections found in the upper respiratory tract. Environmental organisms. Have been associated with warming drying off tubes in contaminated water prior to infusion. Very unusual and cows can die from a toxaemia.

Salmonella

Has possible human health implications.

Serratia species

Can cause mastitis in both dry and lactating cows. Found in soil and plants. It can grow in chlorhexidine disinfectants. Chronic infections which respond very poorly to treatment.

Streptococcus faecalis

Present in faeces.

Causes of clinical mastitis

Table comparing some causes of clinical mastitis

	E. coli	Klebsiella	Pseudomonas	Strep uberis	Strep aureus	Bacillus	Myco-plasma	Prototheca	Yeasts, fungi
Organism	BACTERIA							ALGAE	YEAST
Associated with		Wood products	Contaminated water	Straw bedding				Contaminated water	Poor hygiene
Properties									
Contagious				Y	Y		Y		
Environmental	Y	Y	Y	Y		Y		Y	Y
Type of mastitis									
Clinical	Y	Y	Y	Y	Y	Y	Y	Y	Y
Subclinical				Y	Y		Y	Y	
Dry period infections	Y			Y					
Toxic cases	Y	Y	Y						
Gangrenous mastitis					Y	Y			
Repeat cases of mastitis				Y	Y				
Response to treatment	Good		Average		Poor	Average	Poor		
Death	Y	Y	Y		Y	Y			

Collecting a sterile milk sample

Equipment for sterile sampling

It is essential to ensure that milk samples are collected in a sterile manner. It is easy to contaminate samples and end up with meaningless results. This is costly and time wasting.

Equipment required for sterile sampling includes:

- Sterile sample bottles. Do not use sample pots that contain preservative tablets that are used for collecting cell count samples
- Indelible or waterproof marker pen
- Disposable gloves
- Predip or disinfectant solution
- Paper towels
- Cotton wool
- Surgical spirit

BEST PRACTICE

- Have a dedicated CMT and milk sampling box

TOP TIPS

- Never rush sterile milk sample collection
- When sampling high cell count cows, just sample the high CMT quarters

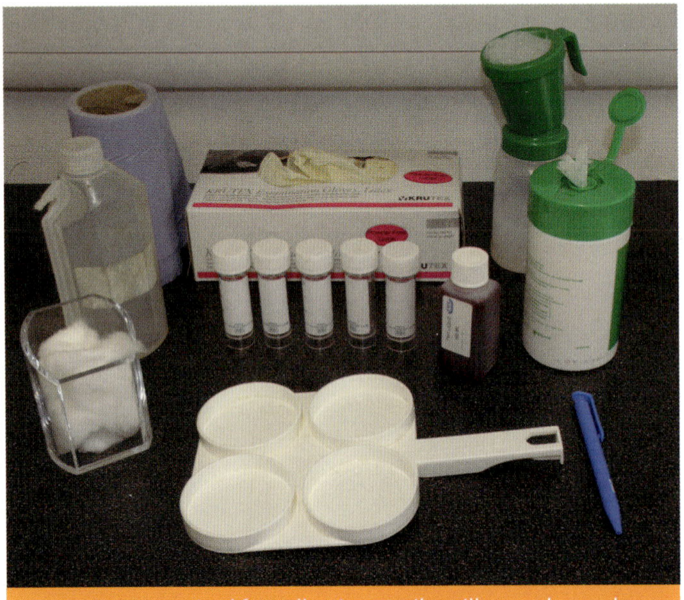

Equipment required for collecting sterile milk samples and CMT testing

From left to right; cotton wool, surgical spirit, paper towel, gloves, sterile sample pots, CMT paddle, CMT reagent, predip disinfectant wipes (which could be used in place of surgical spirit and cotton wool) and permanent marker.

Collecting a sterile milk sample

How to collect a sterile milk sample

1. **Label the bottle.** Write the farm name, cow number, quarter and date on the label using an indelible marker pen. Some people like to mark the side and top of the pot.

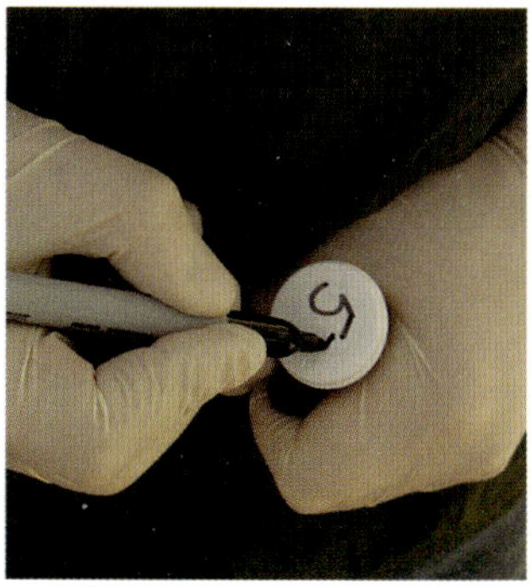

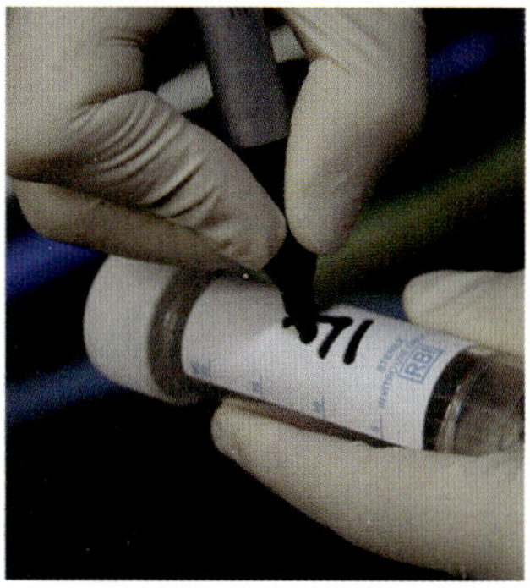

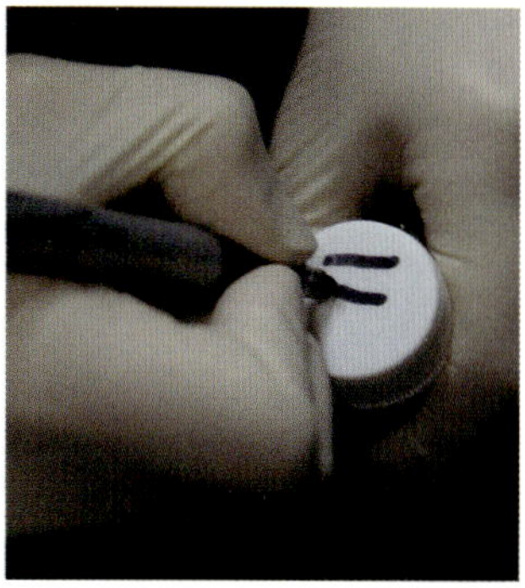

2. **Restrain the cow.** Ensure the cow is well restrained so she can't move around or kick the operator. It might be helpful to have another person hold the cow's tail up.

3. **Put on the disposable gloves.** This is important to avoid contaminating the milk sample.

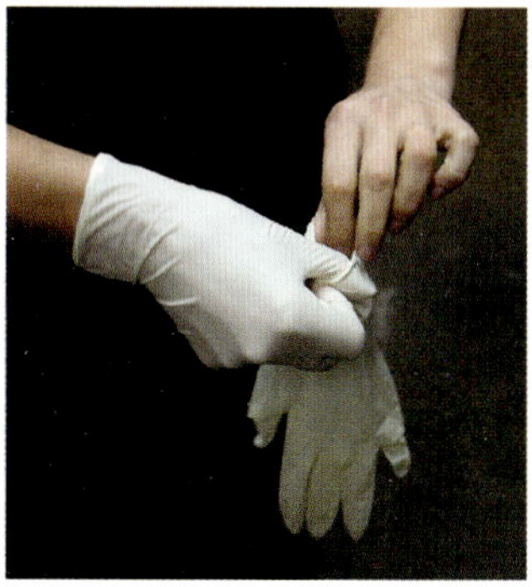

4. **Strip out the teat.** Strip out three or four squirts of milk to remove the foremilk to remove any bacteria that might have entered the teat between milking.

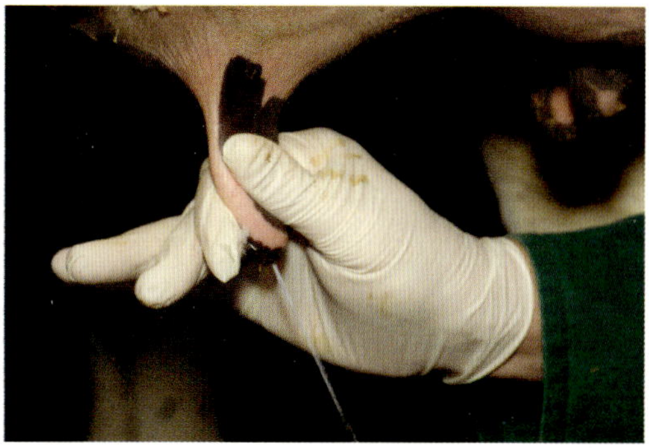

Collecting a sterile milk sample

How to collect a sterile milk sample

5. **Clean the teat.** (a) Predip the teat or teats to be sampled. (b) Allow a contact time of 30-60 seconds and (c) wipe dry with clean paper towel. If predip is not available, wash the teat in running water and wipe dry with paper towel. Take care not to wash the base of the udder as water could drain down onto the cleaned teat and contaminate the sample. Now you have a clean dry teat.

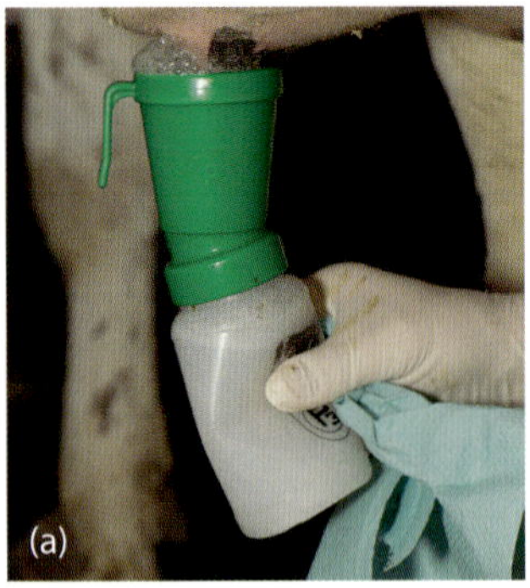
(a)

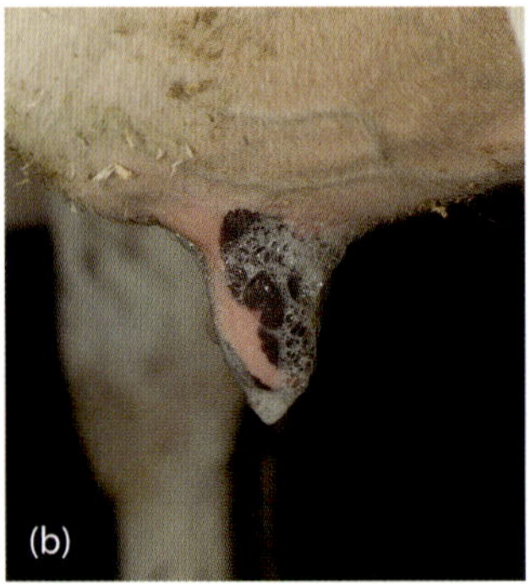
(b)

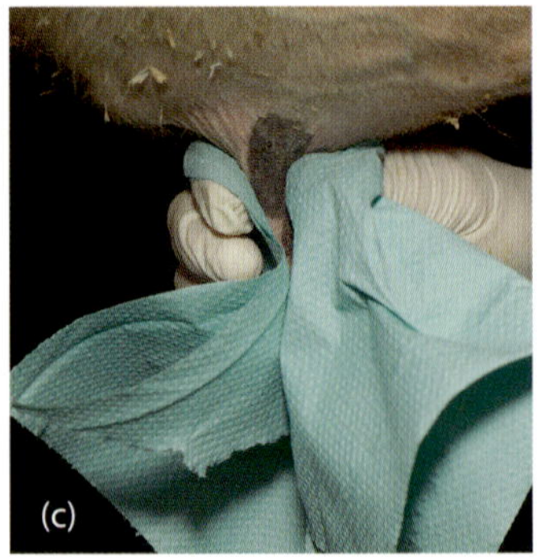
(c)

6. **Disinfect the teat end.** (a) Disinfect the teat end using cotton wool soaked in surgical spirit or the disinfectant wipes supplied with intramammary tubes. Scrub the teat and thoroughly. (b) If there is dirt on the cotton wool after wiping clean, repeat the process so that the cotton wool is always clean prior to sample collection (c). If you are sampling more than one teat, always disinfect the furthest teat first and then the nearest.

WATCH OUT!

- Wipes for preparing teats before milking are not adequate to disinfect teat ends

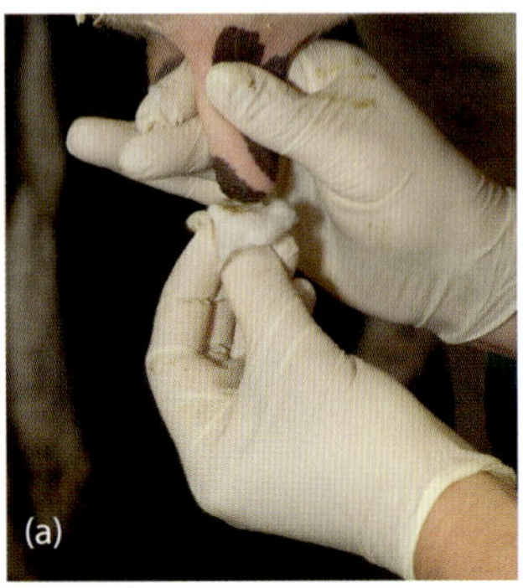
(a)

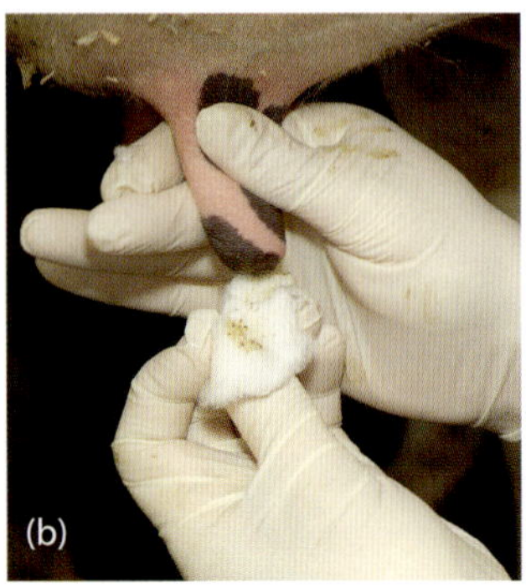
(b)

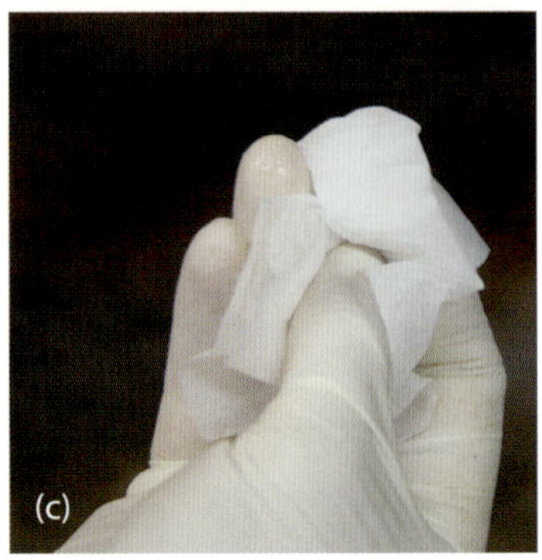
(c)

Collecting a sterile milk sample

How to collect a sterile milk sample

7. **Sample collection.** (a) Unscrew the top of the sterile sample pot and hold the pot upside down so that contamination does not fall into the pot. (b) Hold the teat, ensuring that your hand does not come into contact with the disinfected teat. Turn the pot so that it is at a 45° angle, about 3-5cm away. Squirt one good flow of milk into the sample pot. If you are collecting samples from more than one quarter into the same pot, then sample the closest teats first and the furthest last to minimise the risk of contamination (c). Replace the top on the sample pot.

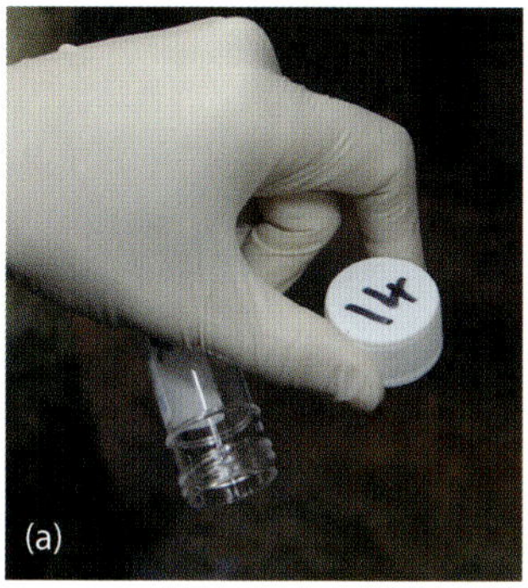
(a)

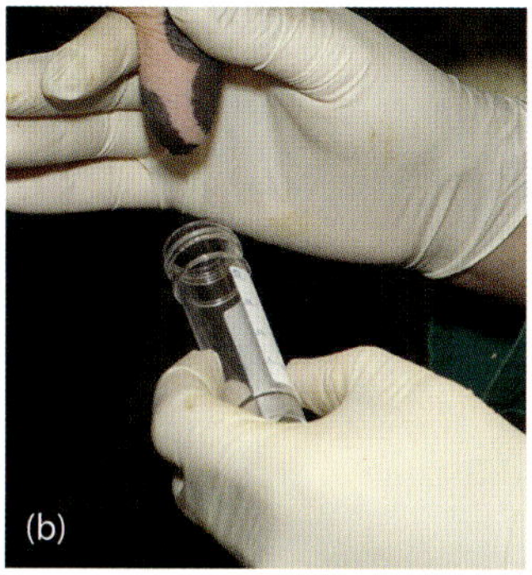
(b)

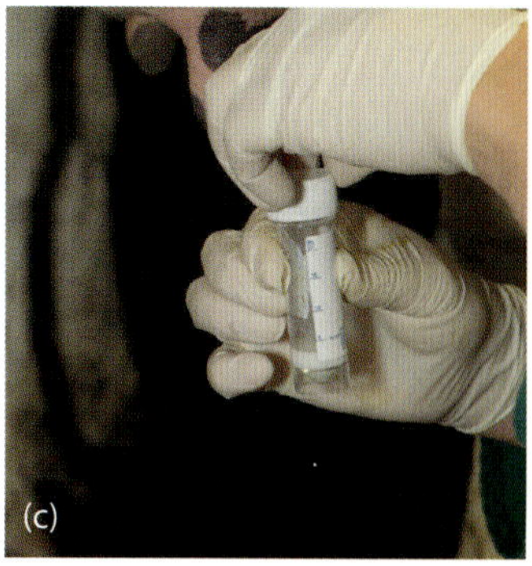
(c)

8. **Storage.** It is important that all milk samples are kept as cold as possible, ideally 4°C or less, from sample collection until the laboratory. These samples are stored in the fridge overnight.

9. **Information.** The more information that can be given to the testing laboratory about the types of samples submitted, such as clinical or high cell count, the better as they can offer enhanced testing for certain bacteria and this should help to improve the overall detection rate from samples.

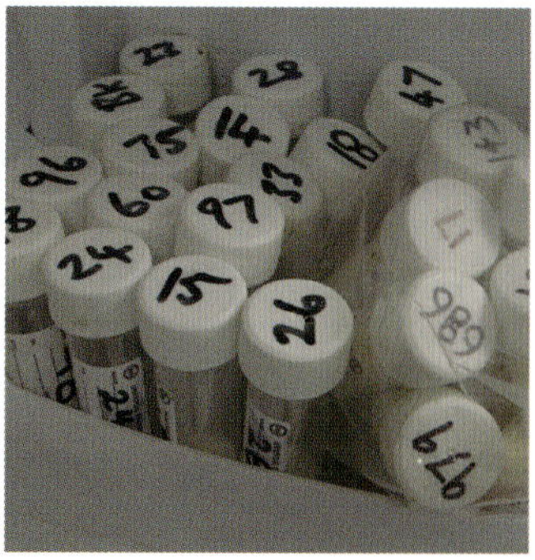

Collecting a sterile milk sample

How to collect a sterile milk sample

10. **Sending samples to the laboratory.** Samples need to be sent so that they will arrive under 4°C , ideally the following day. Samples should be packed in a polystyrene container (a) with ice packs on top of the samples (b). If there is any dead space in the box this will allow the temperature to rise. Fill all dead space to maintain a low temperature (c).

> **TOP TIP**
>
> - Make sure your fridge is working correctly to keep samples cool

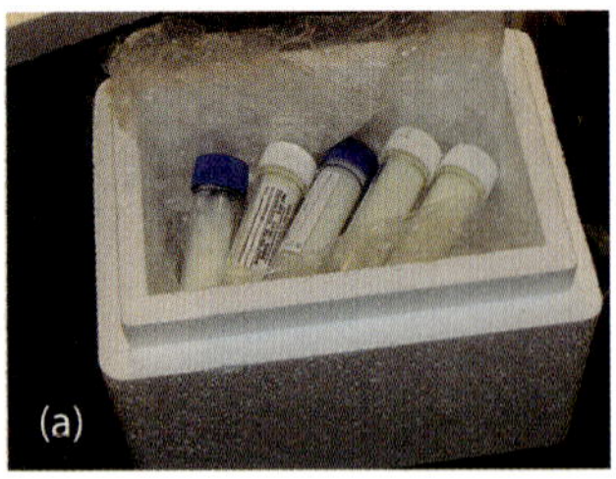
(a)

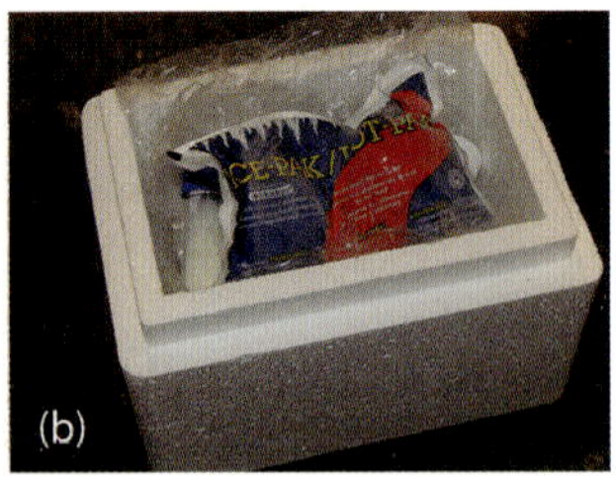
(b)

(c)

It is very important that the lid fits correctly (d) and that it is securely taped (e) so that the inner contents are sealed to avoid any temperature loss.

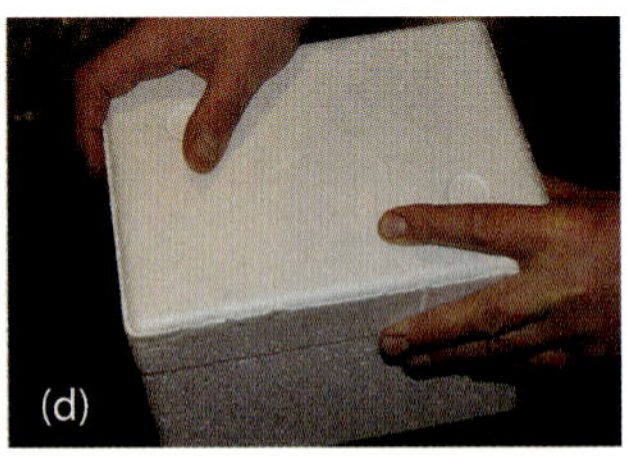
(d)

(e)

Ideally samples will be dispatched guaranteed next day delivery. If samples get delayed or are posted just before the weekend they will get warm and it is possible that the results will be contaminated.

Environment

How to control clinical mastitis

5

Environmental bacteria

Minimising the risk

The environment is very important from a mastitis point of view as this is the source of the majority of clinical mastitis infections. The aim has to be to keep the udders and teats as clean as possible. If this can be achieved then the risk of clinical mastitis from environmental bacteria will be reduced. There are many herds which have excellent housing and well managed beds which have very low levels of clinical mastitis.

A good environment reduces stress levels and promotes happy and healthy cows which will maximise immunity and help fight off any infections that might be encountered. Adequate housing and comfortable beds will allow maximum lying times and reduce lameness which also helps to reduce mastitis.

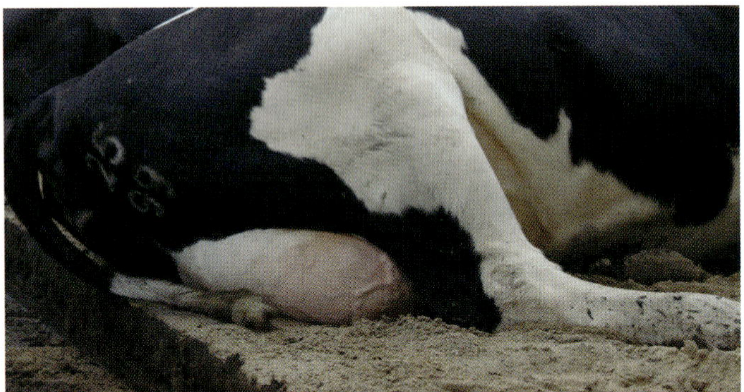

Spotless cows on a deep sand cubicle with very low levels of mastitis

Cubicles

Maximising cow lying times

Cubicles are often called freestalls and are the most common form of cow accommodation used. Care and consideration needs to be given to today's dairy cows which are much bigger than their predecessors.

Cubicles should be designed so that they are comfortable to maximise lying times. A cow will lie down and rest for up to 14 hours per day if she has a comfortable bed. If cows are lying on concrete or are standing in the cubicles this suggests that they are not comfortable.

Every cow lying on this row of cubicles tells us that these are very comfortable beds

Lots of cows standing in these cubicles. They want to lie down but they are reluctant to do so because there is a solid wall in front making getting up difficult

Several cows in this herd were lying on the concrete yards as the cubicles were not comfortable

The backs of these cubicles are very dirty indicating a problem with the design or management

Cubicles

Ensuring cows can lie in the correct position

You need to make sure that the cubicle design allows the cow to lie straight so that when she passes manure this ends up in the passageway. You do not want cows lying crossways in cubicles as this will result in dirty cubicles, mastitis and will increase the risk of cows getting injured.

A brisket board is placed towards the front of the cubicle bed to stop cows lying too far forwards. It is about 68 inches or 172 centimetres from the rear lip. This dimension will vary according to the breed and size of the cows.

A brisket board stops cows lying too far forwards. This cow is resting her right knee on the brisket board

These cows are correctly aligned in the cubicles with their backsides just over the passageway

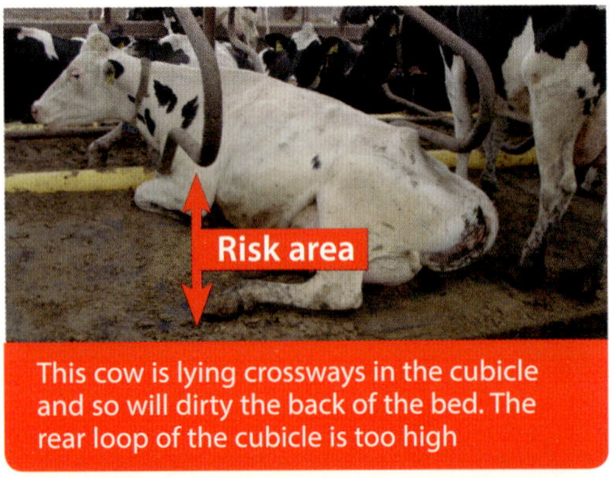

This cow is lying crossways in the cubicle and so will dirty the back of the bed. The rear loop of the cubicle is too high

These cows are lying too far forwards and across the cubicles due to poor design and lack of a brisket board

These broken cubicles should be fixed to maximise the number of available cubicles. It also ensures that cows lie correctly

Cubicles
Lunge space

Cubicles need adequate lunge space at the front of the cubicle. When cows get up they lunge forwards between 70 to 100 centimetres and they need to be able to do this easily. If a cow has a solid wall in front of her then she will find it difficult to get up. This will decrease cubicle acceptance and increase standing times which will have a negative effect on mastitis, lameness and overall health.

These popular cubicles are in a single row and have lots of lunge space and so getting up is easy

A double row of cubicles. Note how much space there is between cows for lunging forwards and to give cows space

These cows have lots of clean loafing space. There needs to be adequate loafing area so that cows which are bulling can exhibit this behaviour without disturbing others.

Cubicles

Passageways

Wide passageways help to keep cows clean as they will have less manure in a given area and this will minimise how much manure gets walked up into the cubicles. If you have narrow passageways then when a cow backs out of one cubicle she could pass manure onto the back of the cubicle behind. Passageways with incorrect drainage also pose problems with dirty cubicles.

Wide passageways help keep cubicle beds clean as cows walk less manure onto the beds

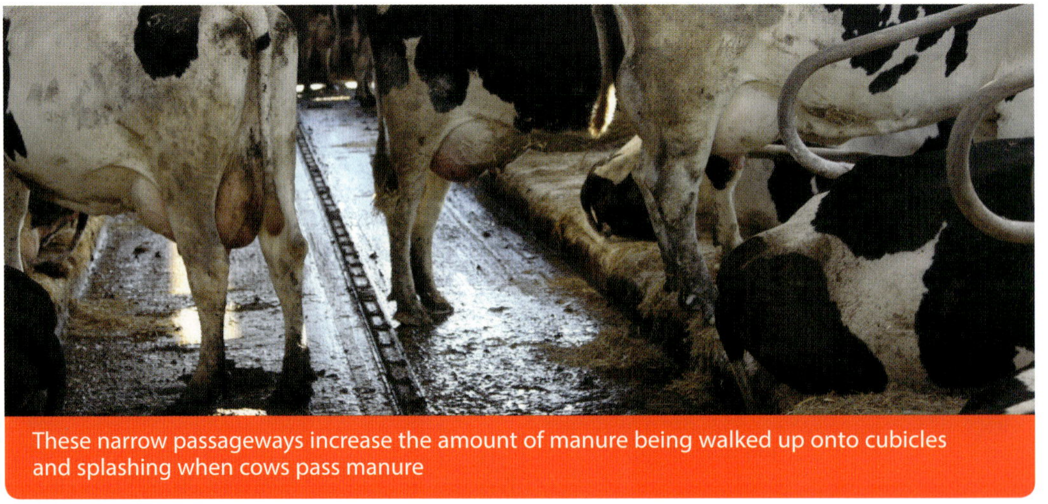

These narrow passageways increase the amount of manure being walked up onto cubicles and splashing when cows pass manure

There should never be dead ends and plenty of cross passageways so that cows do not feel threatened and can escape from dominant cows. This is especially important for heifers.

Cubicles

Maintaining a clean environment

The most important part of the cubicle from a mastitis point of view is the area where the udder lies. This needs to be kept as clean as possible. Most farmers scrape the backs of these beds off when the cows go to the milking parlour. These beds can either be scraped off manually or using a mechanical device.

The backs of cubicle beds must be scraped off and cleaned when cows go to the parlour to be milked

The backs of these beds have been scraped off. There was only one dirty cubicle which is excellent

Cubicles should be bedded up at least daily. This might be carried out manually or using a mechanical device. More big herds use mechanical ways of bedding up which can do an excellent job and are labour saving. People who bed up once a day will pull some of the clean bedding from the front of the cubicle to the rear so the cow has a clean and dry area around the udder.

Some people bed up less frequently either to try and save money on bedding or maybe don't have enough time. There will be an increase in mastitis if the udder and teats get dirty.

Cow comfort and preference

You need more cubicles than cows and the recommendation is to have between 5 and 10% more cubicles than cows. Cows like choice. Some cubicles will be more favoured than others. Heifers will not sit next to or opposite a dominant cow. Stocking density is really important especially in expanding herds where they might not have enough cubicles but want extra milk. This might come at considerable cost.

If you have too many cows for your accommodation then it is advisable to dry off stale cows to reduce stocking density. If this is not possible then overstock those in late lactation who will have a lower risk of mastitis.

The cubicle base has a significant impact on cow comfort and preference. This is often seen when a herd builds new cubicles with a different design. Cows quickly work out which are the most comfortable. The table below shows that a rubber filled mattress is by far the most comfortable while bare concrete which is cold and hard will be rejected by cows. There can be a significant difference in the quality and comfort of mattresses and rubber mats.

Cow preference for cubicle base

Base of cubicle	Percent of cubicles occupied
Rubber filled mattress	89%
Deep bedded sand	79%
Rubber mat	65%
Bare concrete	39%

Rubber mats and mattresses

Increasing cow comfort

Many farmers have installed rubber mats or mattresses on concrete cubicles to increase cow comfort. There are a whole range of mats and mattresses available. These greatly increase cow comfort and provides insulation from cold concrete during the winter time. They have a soft flexible surface which adds to comfort.

There should always be some bedding put on beds to avoid friction damage to the hocks and to ensure that the beds are clean and dry. This is important as cows are wet creatures and you will have moisture on mattresses which could encourage bacterial growth without bedding. Adequate bedding is also important in the event of milk leakage which can occur.

Measuring cubicles to double check dimensions. The wall in the front will reduce cubicle acceptance due to limited lunge space

This cow is leaking milk on a very dirty bed and so the risk of mastitis will be high

These mats will not be comfortable for cows. Note there is no forward lunging space. Expect low usage by cows

The sand does not stay very well on these wet mattresses and wet beds increase mastitis risk

You do not want pooling of any liquid on the cubicle bed

It is important that mats or mattresses are not smooth and slippery. You want to the cow to have good grip so that she has secure footing to get up. This will help to increase cubicle usage. Mattresses have an individual bed for each cubicle. The problem with these is that over time some can become pitted and allow pooling of liquids. If this pitting occurs around the area of the udder, there can be dirty liquid which pools and increases mastitis risk.

Straw yards

Keeping the udder and teats clean

Straw yards are often called loose yards or straw courts. These are large areas of deep bedded straw and cows can lie wherever they choose. Cows love well bedded straw yards as they are very comfortable.

Cows pass manure anywhere in a straw yard and so the challenge in these systems is trying to keep the udder and teats clean. There is a great risk of *Strep uberis* mastitis in these yards. Clinical mastitis levels tend to be much higher in straw yards compared to well managed cubicles because you cannot stop cows lying on manure.

A cow that lies on this piece of manure has a high risk of getting *Strep uberis* mastitis

These cows have plenty of space in a well bedded yard

This shows a step which separates the concrete and bedded area of a straw yard

Straw chopper with bale

Straw chopper with bedding up

Straw yards

Maintenance

Straw yards take a lot of management and time to maintain. They must be bedded up at least daily with plenty of clean dry straw. Ideally, you should bed up every time cows go to be milked so that they always return to a clean and dry bed.

The amount of straw used in these systems is very high and so this can be a very expensive form of housing compared to cubicles. The labour requirements are also greater.

Straw yards need to be cleaned out at least every 2 to 3 weeks, maybe more frequently if the temperatures are high. A cow produces about 50 litres of liquid each day and much of this will end up in the straw. You then have a mix of liquid, manure and heat which provides everything you need to grow bacteria. If you dig down a few inches into a straw yard and place your hand just above the bedding you will feel how hot it is. The bed is about 40°C which is the ideal temperature to grow bacteria! No wonder there is so much mastitis in these yards.

If you dig down into the straw bed you can feel the heat from the straw. You can see the steam rising from this straw bed

Cleaning out straw yard. Note the steam rising from the hot dirty bedding, an ideal way to grow *Strep uberis*

Straw yards
Design

The design of the yard is important. They should be wide and shallow and not narrow and deep. This helps to avoid cows walking over large parts of the beds and contaminating it. This helps to keep the straw in the bedded area and to reduce the amount of straw contamination.

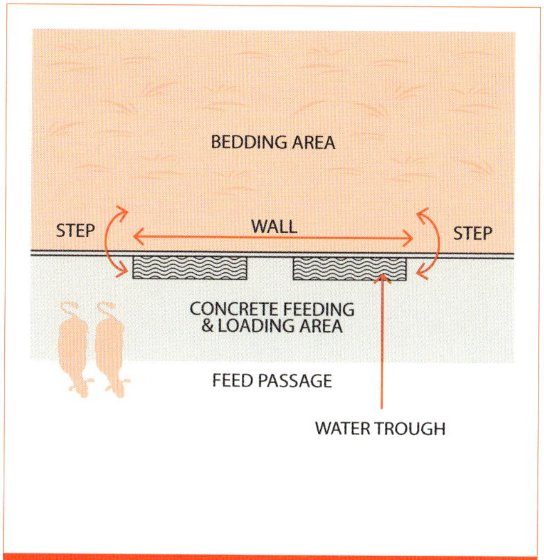

Well designed straw yard

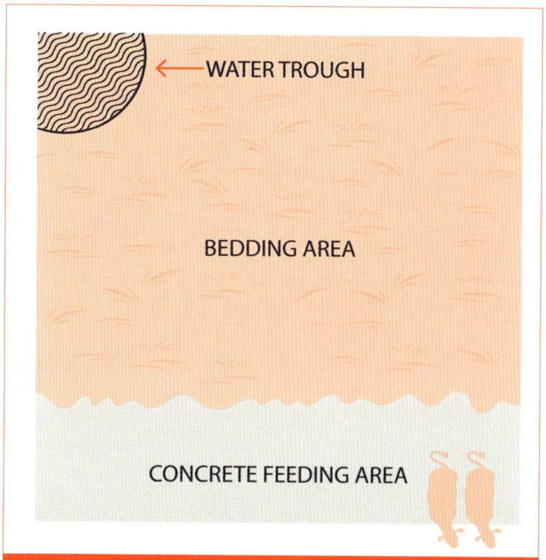

Poorly designed straw yard

This straw yard is poorly designed and badly bedded. These yards should be on a free draining base to avoid pooling of liquids

Straw yards

Maintaining a clean environment

Some herds bed up twice a day in the straw yards to ensure that the cows are kept very clean and the risk of the udder coming into contact with faecal matter is minimised.

This shed is bedded up twice a week and allows a build-up of dirty straw. The cows were not clean

There is no barrier between the bedded and concrete areas and straw gets dragged everywhere

This straw yard is very overstocked with a high risk of *Strep uberis*

This straw yard is very dirty. The owner cut back on straw use to save money

Straw yards
Water trough positioning

Water troughs should never be situated in the bedded area. They should be located so that when cows drink and pass manure this always falls onto a concrete area. Some yards have water troughs situated on the bedded area and the straw around it becomes very dirty. This also reduces the amount of lying area.

You need to allow plenty of space for cows in straw yards. The recommendations have changed over the years. Originally, it was thought you needed relatively small amounts of space. For a 750 Kilo cow allow about 9 to 10 square metres of lying space and 3 square metres of loafing and feed area.

Correctly positioned water trough. Cows can only drink from the concrete loafing area

Incorrectly situated water trough. Cows can drink from the bedded area which will become contaminated and wet

Concrete areas

Passageways, parlour, feed, exit areas and collecting yards

It is important that all concrete areas and passageways are kept as clean as possible. Some passageways are narrow and get heavily contaminated after milking. Cows can end up splashing manure on their udder and teats on the way to and from the parlour.

Splashing will be a greater risk with frisky or nervous cows which have a tendency to run. If they splash manure on the open teat end then there is a risk of mastitis bacteria entering the udder.

All concrete areas should be scraped every time that cows leave the building to go to be milked. It is important that attention is paid to cross passageways and the areas just behind the cubicle step to keep them clean.

Clean and well scraped loafing and feed areas will help keep cows clean

The cross passages here have been well scraped before the cows return to the shed

Automatic scrapers can be useful in keeping passageways clean but they need to be run frequently

These cows have to stand in deep liquid manure to feed. The mastitis and lameness risk here were very high

Lots of space at the front of the parlour so cows can leave at their leisure without rushing

This cross passageway has not been cleaned

Concrete areas

Passageways, parlour, feed, exit areas and collecting yards

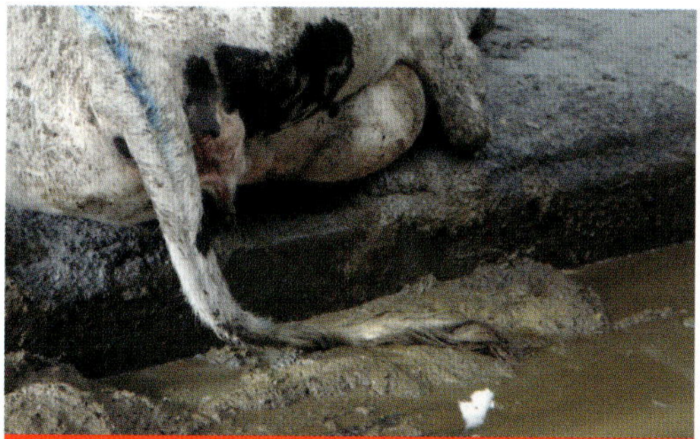

A long tail in a very dirty passageway will splash manure onto the udder and rear teats and increase mastitis

This farmer has not scraped all the concrete areas properly.

Tip-over water troughs help clean cross passageways whilst ensuring that cows always have clean and fresh water

These cows are in a well scraped and ventilated area

Yards are being scraped onto slats which drop down into the slurry system

These box scrapers are very efficient as no manure flows out of the sides of the scrapers

Flood gate washing

Flood wash system

Some cubicle houses use a flood wash system instead of scraping to wash the manure down the passageway using large amounts of water. Often this is waste water that is far from clean. This flooding has been known to take place when cows are standing in the passageway. The danger is from splashing of these liquids onto the udder and teats.

This shows the power of flood gate irrigation which comes from the ground. This is using recycled water which contains lots of faecal matter

Flood gate washing when cows are present can cause splashing on teats and increase mastitis

Foot baths

Design considerations

More and more farms are now foot bathing cows after every milking as they exit the parlour. This is a very effective way to help reduce lameness but care and consideration needs to be given to the design of any footbath.

Ideally you would have a wide footbath so that two cows can pass each other. Cows do have a habit of lingering and looking, after all they are curious animals.

This wide footbath allows cows to overtake without being rushed and reduces the risk of splashing

The footbath should not be overfilled and needs to be just above the level of the fetlock and topped up as necessary. If the footbath is very deep then there is a risk of splashing liquids up against the open teat canals.

This footbath is situated at the exit to the parlour and is far too deep. There was faecal matter splashing on open teats contributing to mastitis

The author visited a farm where the cows stepped down into a footbath and as they did some teats were rubbing the concrete step into the footbath. Not only would this have been painful for the cow, but the teats got heavily contaminated. Some cows didn't like the step and so jumped in splashing muck all over the udder and teats. Levels of mastitis dropped once the footbath was moved.

Pasture

Risks to mastitis

It is assumed that pastures are clean and not a significant risk to mastitis. This is far from the case. Cows can become very dirty when at grass and can introduce *Strep uberis* infections onto pastures and mastitis outbreaks can occur.

Cows will often lie in the same area, especially if it is hot or wet when they are seeking shelter. In some countries cows calve in calving paddocks and if these are overstocked the risk of environmental mastitis will increase significantly.

Consideration should also be given to the tracks or paths out to the field. Sometimes when the weather turns wet these cows have to wade through deep mud and manure to reach the field, increasing the risk of mastitis occurring.

Cows enjoy the space and freedom of being at grass

The cow track is to the right of the drive for the tractors and farm vehicles

These cows can walk down a wide and clean track to graze

Corrals

Risks to mastitis

Cows may live in corrals in some countries which have a dry climate for the majority of the year. Many of these corrals are just fenced off paddocks with areas for silage and other feeds. When conditions are dry these work very well and cows can remain very clean. If the climate is hot then liquids evaporate quickly.

However, when conditions turn very wet the corral can turn to mud and this can greatly increase the risk of mastitis. Some cows like to lie in this mud with their udder totally submerged as it's cool and comfortable. The risk of mastitis in these situations is exceptionally high as the teat end is bathed and constantly challenged by millions of environmental bacteria.

The graph below shows the mastitis incidence throughout the year in a 400 cow African dairy herd. Note the massive levels of mastitis in December, January and February which coincided with the rainy season that year.

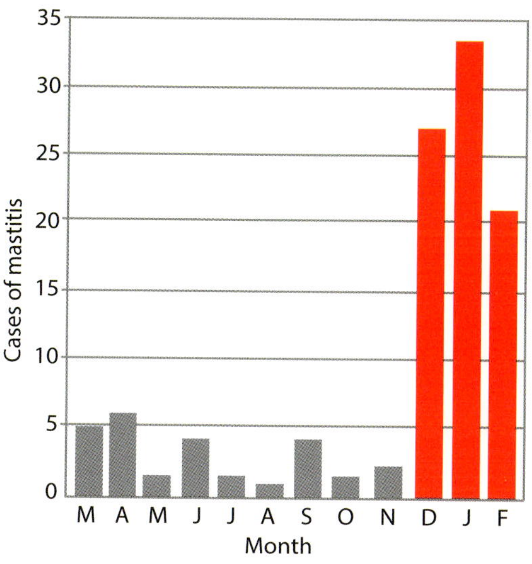

Mastitis incidence throughout the year

This cow is very happy buried in mud in the hot season in Africa. The risk of clinical mastitis will be massive

This bobcat is cleaning off the dirt from this corral and pushing this into a large pile which will be removed

These cows have plenty of space in a well managed corral

It is important to harrow sand corrals daily

Calving pens and paddocks

Minimising the risk of mastitis

Calving areas should always be kept clean to minimise any risk of mastitis and also to ensure that the calf remains healthy. Sometimes these areas are neglected. Remember that the freshly calved cow's immune system is weakened around calving and so there is a much higher risk of mastitis at this time. The same is true for dry yards which are often neglected or bedded up infrequently.

The cow has a large pen to herself and her newly born calf. It's clean and the risk of mastitis will be minimal. It's cleaned out and bedded up between uses

This communal calving pen is overstocked and dirty. There is not enough clean bedding

This cow has calved in a wet and dirty area where there will be a greater risk of mastitis

Bedding materials

Choosing the right bedding materials

Bedding materials can be divided into two categories; organic and inorganic. Organic bedding materials, like straw, have nutrients that encourage bacterial growth and can increase and affect mastitis levels. Inorganic materials, like sand, are best as they are inert as they do not have nutrients that allow bacterial growth.

The choice of bedding materials will depend on a number of factors;

- Availability
- Compatibility with the manure system
- Cost
- Cow comfort
- Mastitis control

Often there is compromise as few bedding materials will meet all the preferred options.

Bacteria require the following factors to grow; nutrients, moisture and temperature. If any of these are missing then growth will be affected. For example, if beds are kept very dry then bacteria will find it difficult to grow. Thought must be given to bacterial growth in bedding materials.

Straw

Straw is generally chopped before put on beds to maximise absorbency. It is important that straw is kept under cover and that only dry straw is used. There is a risk of *Strep uberis* mastitis with straw bedding but this can be minimised provided it is well managed.

If straw is chopped too finely it can blow out of cubicles and so you want a chop length which retains the straw in the cubicle. You generally use about 460 kilos of straw per cubicle for a six month housed period. This compares to 3,660 kilos in a loose yard system for the same period, eight times as much!

One of the disadvantages of straw is that there can be considerable price variation and that it is available at set times of the year. It needs space to be stored under cover. Some people store straw outside and even if it is covered with plastic, a lot of this straw will be spoilt. Once straw gets damp then its ability to absorb water decreases. It can become contaminated with ons and moulds which can mastitis.

Straw correctly stored under cover

These bales are stored outside. Note the mould and waste on the left bale. These will not absorb much moisture

Bedding materials
Sand

Sand is an excellent bedding material as it is inert; bacteria do not grow in sand. Ideally cows will be on a deep sand bed. There are many different types of sand available and it is important that washed sand is used as this removes any organic contamination. Sand should be stored in a dry area so that dry sand is put on beds. Sand is now commonly used in cubicle beds and farmers who manage this well have seen very significant reductions in mastitis.

This sand is stored outside and will be wet when put on the beds which is far from ideal

Correct storage of sand. Washed builders sand stored under cover

Cows love these sand cubicles which are all full. Note that cows push sand out of the cubicles and into the passageways

Sand beds should be on a free draining base so that liquids can drain. You need to remove any manure soiling from the backs of the beds regularly and top up generously with fresh sand.

Cows like deep sand beds as they are comfortable and give great grip when they have to get up. This is seen with lame cows who can often struggle in the likes of cubicles with rubber mats which can be slippy.

Bedding materials
Sand

It's advised to rake the top of the cubicles daily to help with drainage. Rake to a maximum depth of 7 centimetres or three inches. If you rake more deeply you can drag up the contaminated sand at the bottom and this can increase mastitis. If the sand becomes hard, compacted or dark then it should be removed and replaced.

This rake is useful for maintaining good drainage in sand. Only rake the top 7cm or 3"

One of the disadvantages of sand bedding is its effect on slurry or manure systems. It sinks to the bottom of slurry pits and can wear through the bearings of slurry handling systems. Most farmers are happy to incur these extra costs as they see significant benefits from sand beds.

Some people have put old car tyres in the base of sand cubicles to reduce sand use. Tyres tend to rise to the top and make it difficult to rake and manage the beds. They are uncomfortable for cows

Bedding materials

Sawdust and wood shavings

There are two types of sawdust; kiln dried and fresh. Kiln dried sawdust will contain about 5% moisture while fresh sawdust can contain as much as 30 to 35% moisture. Kiln dried sawdust is the only suitable sawdust to use and is a modified by-product. Many farmers buy dried sawdust in bulk and it is essential that this is stored so that it remains dry.

Type of sawdust	Dry matter
Kiln dried	95%
Fresh	60-70%

Fresh sawdust stored under cover to avoid spoilage

Sawdust is finely ground with a large surface area and so has great absorbency properties. *Klebsiella* like any of the wood bedding materials. It is naturally found in wood and is essential for the decaying process. The problem is that fermentation commonly occurs when you have fresh sawdust. Put your arm into a pile of fresh sawdust and if it feels very hot you know that it is fermenting. This will greatly increase *Klebsiella* mastitis.

If sawdust feels hot it is fermenting and there is a high risk of *Klebsiella* mastitis

Some farmers try to save money, for example when milk price or profits fall, and herds have moved from kiln dried to fresh sawdust only to end up with outbreaks of clinical mastitis. The graph below shows the changes to mastitis in a herd which was housed all year round where the only change to management was the type of sawdust used. You can see that when fresh sawdust was used, as shown by the red columns, mastitis levels doubled and then dropped back when the farm changed back to kiln dried sawdust, shown as the grey columns.

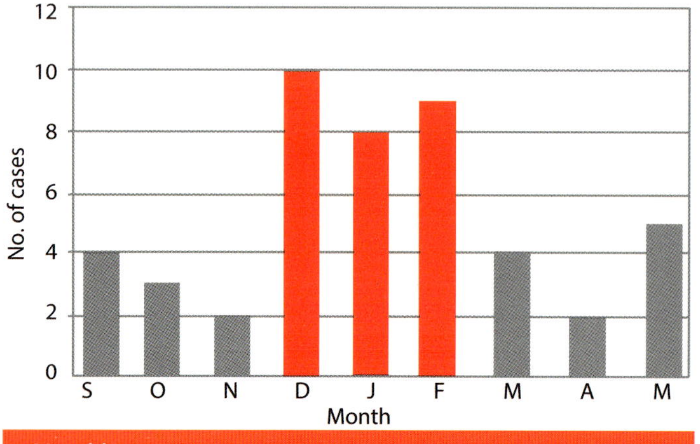

Monthly mastitis cases

Fresh sawdust should never be used for bedding cows even if it is offered and delivered free of charge.

Bedding materials
Paper and paper based products

These are by products of recycling. Shredded paper has been used by some but it is not very absorbent and so has limited value. It also has a habit of sticking to cows.

There are some manufactured paper bedding products such as Envirobed which is made specifically for use in bedding and has 95% dry matter, the same as kiln dried sawdust. These are paper pellets and quite absorbent.

They are available all year round and can be purchased as required compared to the likes of straw. These products all break down in the slurry system.

Envirobed is made from recycled paper

Bedding up with Envirobed which can be quite dusty

Bedding materials

Recycled manure solids

Recycled manure solids, often referred to as RMS, are used to bed cows as you are recycling cow manure and saving on other bedding material costs. They can be used as deep pack bedding such as a loose yard system and as bedding for cubicles. This type of bedding is very comfortable to the cow. In England this form of bedding is illegal to use but the Government authorities have said that they will not prosecute farmers who use this for cows.

Recycled manure solids have become a very attractive proposition for the larger herd which would otherwise have significant bedding costs. However, many have said that this presents a poor image for dairy farming where cows are lying on their own faeces. Some of the larger herds are getting a four year payback from the capital cost for installing the separator to produce recycled manure solids.

Recycled manure solids typically contains about 30-35% dry matter and so while they are referred to as solids, this is not the case. They are more liquid than solid. In arid areas the dry matter will be higher.

Management of recycled manure solids is critical and must be used as soon as they are produced. Old recycled solids must not be used as they will heat up and ferment. Cubicles must be bedded up daily with any dirt being removed beforehand.

Problems can occur in areas where there is high humidity where the moisture content of the recycled solids is even higher. If cows are kept in overstocked yards or cubicle housing this can increase the risk of mastitis. The bacterial load of this bedding is very high, especially Klebsiella. Herds should predip cows to ensure that teats are clean and dry before the milking unit is attached.

There are many herds which do manage these very well without any increase in mastitis, while others have run into mastitis problems. It requires great attention to details and careful management.

A separator removes some of the moisture from manure

Recycled manure solids fall from the separator above and form a pile underneath. These must always be kept covered to avoid getting damp

This herd uses recycled manure solids on top of sand bedding

Recycled manure solids actually contain about 65% moisture

Bedding materials

Other recycled products

A range of other products have been used for bedding. These have included ash from power stations, ground plasterboard and chipboard. Companies are always looking for any way to get rid of this waste product to avoid having to pay disposal costs. While many of these can appear cheap, they are often unsuitable as bedding.

This recycled wood chip will have very poor absorbency and be abrasive, reducing cow comfort

Drying agents

Drying agents are used to help absorb moisture to keep the area under the udder dry. There are a large range of products available such as hydrated lime or commercially formulated products specifically for this purpose. These products can prove very useful, especially in sheds where there is poor ventilation. Any drying agents should be put on the clean cubicle base and then covered with the bedding material. These should not come into direct contact with teats as they can dry the skin of teats and make them rough.

Some farmers mix the drying agent with the bedding material such as sawdust. The intention here is good, but the problem is that the drying agent will come into direct contact with the teat, which is not recommended. This is often carried out to save labour as you only need to apply one product to the cubicle. In well ventilated sheds a drying agent should not be necessary or may be used once a week.

You do not need to use expensive commercial products. You would be better off using the money you save on increasing the amount of bedding used. Hydrated lime is a very effective and cheap drying agent.

Hydrated lime is a useful drying agent for cubicle beds. Apply before putting on the bedding

Bedding materials
Frequency of bedding

Fresh bedding should be put on daily for all milking cows. There are some herds which only bed up cubicles once or twice a week. This might be adequate if they put a lot of bedding at the front of the cubicle which they then rake back every time the cow goes into the parlour. In some herds this practice is designed to be labour and cost saving.

For straw yards, where the cow can come into direct contact with faecal material, they need to be bedded up daily with a generous amount of straw.

Automatic bedding dispensers such as this make bedding up easy and very labour efficient

Bedding materials
Volume of bedding

The amount of bedding material used will depend on a wide variety of factors which include ventilation, stocking density, width of passageways in cubicle housing, design of straw yards and faecal consistency.

Some people bed their cows very generously so that they are meticulously clean whereas others will use sparse amount of bedding material to try and save money.

Bedding type and quantity used per cow per day and for a six-month period

Bedding material	Cubicles (kg/cow/day)	Straw yards (kg/cow/day)	Total use for a 6 month period (kg)
Straw		20	3,660
Straw chopped on mat	2.5		460
Sawdust - Kiln dried	1.0		185
Sawdust - Fresh	2.2		400
Sand	16		2.925
Recycled paper products	2.2		400

Assessment of cow accommodation

Udder hygiene scoring

It is important to assess cleanliness of cows, the udder and teats rather than look at the environment. After all, if the udder and teats are clean then the environment is perfect from a mastitis viewpoint. If the cubicles or straw yard look clean and the cows are dirty then you need to work out when and where the cows are getting dirty and then correct the problem.

A really useful scoring system has been devised by the University of Wisconsin and is called the Udder Hygiene Scoring Chart. This is a simple tick box system where you can score your cows when they are in the parlour. You can do this for different groups and see what the differences are. The aim is to have at least 80% of cows with Scores 1 or 2. If there is a high proportion of cows with Scores 3 or 4 you should expect higher levels of mastitis. Visit http://milkquality.wisc.edu

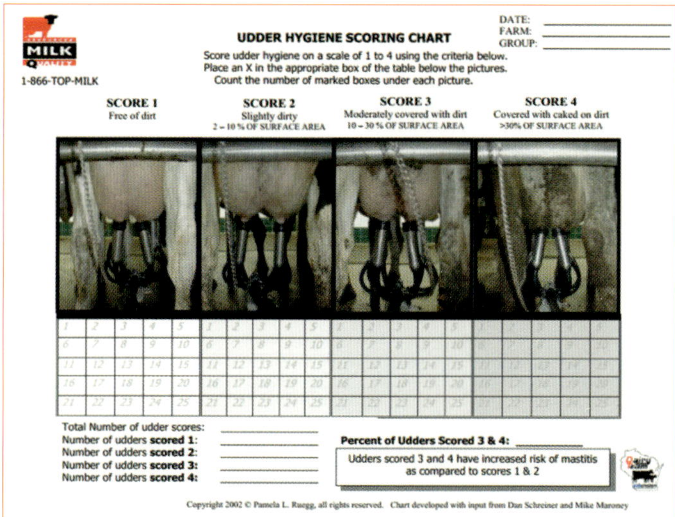

MILK QUALITY

1-866-TOP-MILK

UDDER HYGIENE SCORING CHART

DATE: ______
FARM: ______
GROUP: ______

Score udder hygiene on a scale of 1 to 4 using the criteria below.
Place an X in the appropriate box of the table below the pictures.
Count the number of marked boxes under each picture.

SCORE 1 Free of dirt					SCORE 2 Slightly dirty 2 – 10 % OF SURFACE AREA					SCORE 3 Moderately covered with dirt 10 – 30 % OF SURFACE AREA					SCORE 4 Covered with caked on dirt >30% OF SURFACE AREA				
1	2	3	4	5	1	2	3	4	5	1	2	3	4	5	1	2	3	4	5
6	7	8	9	10	6	7	8	9	10	6	7	8	9	10	6	7	8	9	10
11	12	13	14	15	11	12	13	14	15	11	12	13	14	15	11	12	13	14	15
16	17	18	19	20	16	17	18	19	20	16	17	18	19	20	16	17	18	19	20
21	22	23	24	25	21	22	23	24	25	21	22	23	24	25	21	22	23	24	25

Total Number of udder scores: ______
Number of udders **scored 1**: ______
Number of udders **scored 2**: ______
Number of udders **scored 3**: ______
Number of udders **scored 4**: ______

Percent of Udders Scored 3 & 4: ______

Udders scored 3 and 4 have increased risk of mastitis as compared to scores 1 & 2

 Chart developed with input from Dan Schreiner and Mike Maroney

If you have lots of dirty cows then steps need to be taken to correct this. You can then repeat the scoring exercise a couple of weeks after any changes have been made.

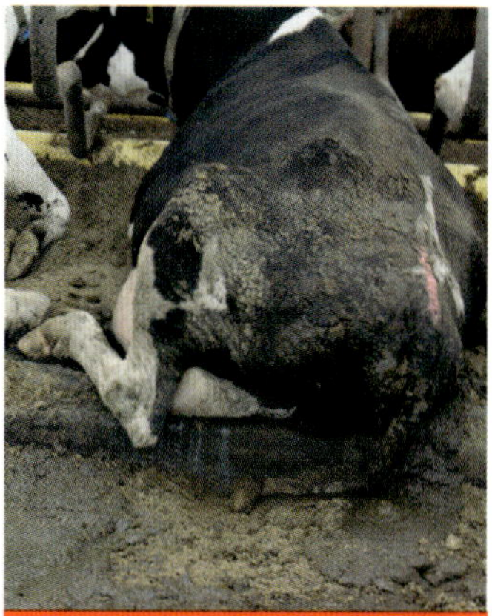

This cow and the base of the cubicle and passageway are all dirty. There is a high risk of clinical mastitis

These two cows are dirty telling us that there is something wrong with the housing management

Ventilation

Housing design

Cows are very wet creatures and good ventilation is required to remove this moisture so the environment can remain as dry as possible. Effective ventilation will help remove excess heat. If you see lots of cobwebs on the roof of a shed it indicates poor ventilation.

This new shed uses raised roof sheets to help with ventilation

Cobwebs in the roof suggest poor ventilation

This cubicle shed has sides that can go up and down according to the climate for effective airflow and comfort

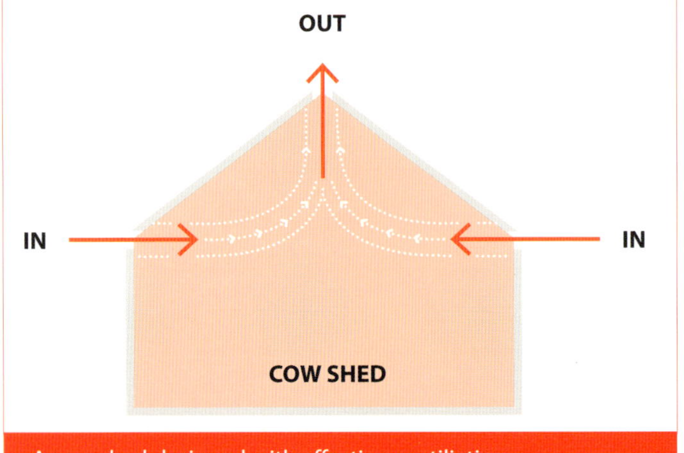

A cow shed designed with effective ventilation

If poor ventilation is present then there is likely to be a build-up of moisture in the environment. This will increase the risk of mastitis organisms growing on the cubicle beds. In addition, poor ventilation will increase the risk of respiratory and other diseases occurring.

Natural ventilation has to be favoured and careful design is needed to ensure that this is effective. Some newly designed cow sheds are over ventilated and there is not enough 'lift' to remove the hot air and moisture.

Housing should have air entering through the side of the buildings and out through an open ridge at the top. Ventilation design is a very complex area and so specialist advice should be taken to ensure that it is effective and that any risk of heat stress is going to be minimised.

Heat stress

Temperature humidity index

Heat stress occurs where the body is unable to maintain normal body temperature. It is a combination of heat and relative humidity, not just temperature. Sometimes cows will be affected by heat stress at temperatures as low as 24°C. The cows comfort zone is between 5°C and 25°C. There is evidence that suggests that the most severe heat stress occurs in short periods and where the cow has not had time to adapt. The temperature humidity index or THI is used to indicate the degree of heat stress and is a combination of temperature and relative humidity.

Heat stress causes body temperature to rise. Cows become less active and water intake will increase significantly. They spend time under the shade and are always looking for cooler areas.

Fans can help cool cows

There are many cows, especially high yielders, which are housed all year round. Some of these cows will be in well designed and ventilated sheds, while others will be in housing with poor ventilation or excessive Perspex sheeting causing a greenhouse effect on hot sunny days.

With heat stress, feed intake can be reduced by as much as 20%. This is the reverse of very cold temperatures when an animal will eat more to generate heat to maintain body temperature.

Milk production can be reduced by between 10 and 25% due to less energy being available for productive functions, as a result of the reduced feed intake. This will be proportional to the severity of the heat stress.

Water intake can increase by up to 50% depending on conditions. Increasing water intakes could prove problematic in some herds which have a poor water supply. The troughs might not be able to fill up fast enough. Problems can also occur if there is inadequate drinking space. Cows which are suffering from heat stress often will linger around water troughs and play with the water.

It's clear that heat stress can have very significant effects on dairy cows and can increase the risk of mastitis.

These cows benefit from a fine mist above the feed area. After a few minutes the mist stops and the fans cut in to cool the cows

On hot days cows will always seek out the shade. Almost all of these cows are in the shade from the midday sun

Heat stress

Steps to minimise heat stress

There is too much perspex sheeting in this shed and it gets very hot in the summer

These cows are sheltering from the blazing midday sun

These cows like the water sprinklers which are situated at the entrance to the collecting yard of the parlour and used on very hot days

A large volume water trough with plenty of clean water for cows to drink as they come out of the parlour

HEAT STRESS CHECKLIST

- Ensure that you have plenty of water tough space, clean water and good water flow rates so the troughs never run dry. You should have a minimum of 10 cms of trough space per cow
- Check for excessive amounts of Perspex sheeting on shed roofs and paint over some if you are getting a greenhouse effect
- Check ventilation levels. You need to make sure that hot air can readily escape from the buildings. Best to use a building ventilation expert for this
- Minimise overcrowding as this will build up heat levels at cow level
- Minimise time in holding pens where there will be poor ventilation
- Consider using sprinklers or water misters. These can prove very effective especially around the feed area. They can provide a very fine mist of water onto the backs of the cows at set times. Evaporation takes places and helps keep cows cool and maintain feed intakes. Using fans along with misters will increase the amount of cooling even more
- Is there adequate shade available? Cows like to stay out of direct sunlight on hot days
- Fans in cattle houses can assist with air movement

Milking Routine

How to control clinical mastitis

6

Milking routine

Introduction

The aim of a good milking routine is to milk cows as swiftly as possible with minimal impact on mastitis. Milk must be extracted in a clean and efficient manner. This means that the milking cluster is attached to a clean dry teat which has been well stimulated to take advantage of the milk let down reflex.

The machine must be well maintained and working efficiently. The cluster should be removed after the cow has been milked out to minimise any teat damage. The cow is then post dipped to help kill any bacteria which have been transferred to the teat during the milking process.

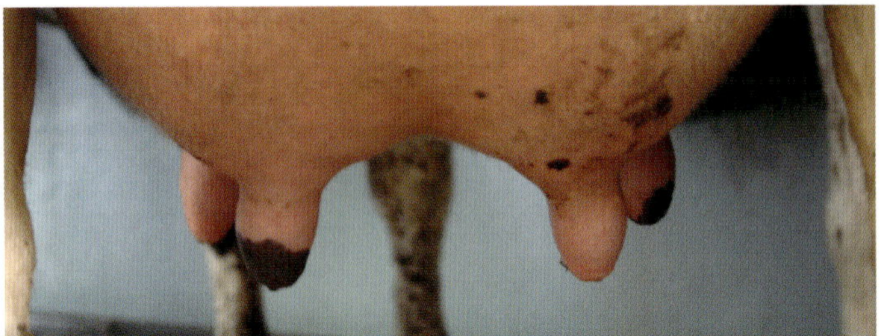

Teats should be clean and dry with maximum milk let down reflex before the milking machine is attached

The clusters must be kept clean throughout milking

A milking parlour is a food factory

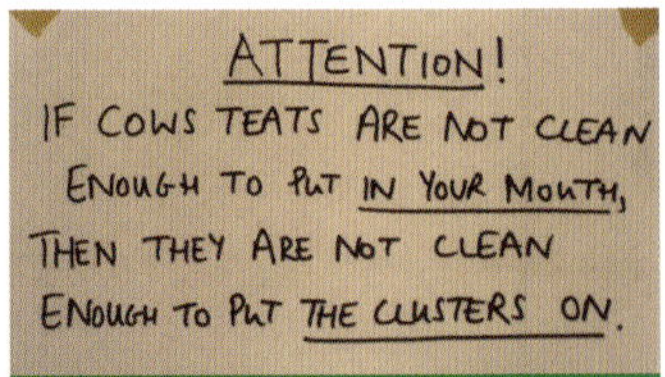

This farmer makes it clear what he expects from his milkers

FACTS

- The milking parlour is a food factory
- The milking machine must be well maintained and working efficiently
- A poor milking routine can increase clinical mastitis, cell counts and Bactoscan

The milking parlour is a food factory. The quality of the milk produced cannot be enhanced or improved. A poor milking routine can result in an increase in clinical mastitis, higher cell counts and Bactoscan or TBC (Total bacteria Count) and result in poor quality milk for which the farmer is paid a lower milk price.

Milker training

A key part of mastitis control

Milkers should be trained so they understand the milking routine and this should be ongoing and updated as necessary. The significance of the milking routine on mastitis and cell counts should be explained so they understand the impact of what they do on mastitis and milk quality.

A written SOP (standard operating procedure) is essential in herds with multiple milkers to ensure that there is consistency in the routine. Herds which have a consistent milking routine can get up to 5% more milk out of the cows than herds which have variable routines. This can occur where there is no agreed way to milk cows and where each milker works to his or her own preference.

Good milker training is a key part of mastitis control. The milkers can make or break mastitis and milk quality and so training should be ongoing.

Milker schools are a great way to ensure that mastitis is minimised

FACT

- Milker training is a key part of mastitis control

GOALS

- Cows should be milked as swiftly as possible with minimum impact on mastitis
- Milk must be extracted in a clean and efficient manner

Milking environment

Happy and content milkers and cows

Milkers should work in a good environment with adequate space to move around and good lighting so they can clearly see the udder and teats. More and more parlours now have rubber mats on the parlour floor for milkers which makes walking up and down the parlour more comfortable and less tiring. This will help milkers do a better job. Milking should not take an excessive amount of time. There are some farms where milkers work for five or six hours twice a day and these people are likely to get tired and cut corners. A well designed parlour will pay dividends in a job well done.

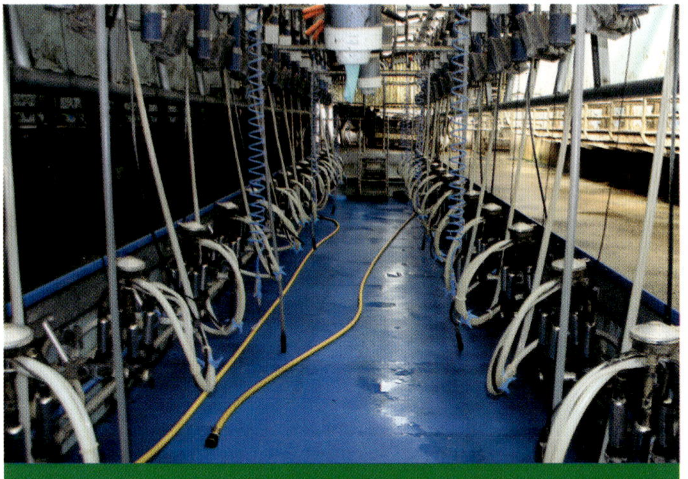

Rubber matting floors make milking more comfortable.

The parlour should be a place where cows and milkers are happy and content. If cows are nervous or frightened then they will pass manure more frequently during milking and the milk let-down reflex can be reduced. This will result in more milk remaining in the udder at the end of milking. These factors can have negative effects on milk quality and mastitis.

Rubber matting on the parlour floor adds to cow comfort and avoids cows slipping down. A large exit area at the front of the parlour will speed up cow flow, as many cows like to linger and look back at the parlour when they have been milked. If parlours have narrow exits these can block cows leaving and slow down milking.

Good light is important and cows should also have good light in the collecting yard. Cows do not like to walk from a dark collecting yard to a bright parlour. This will slow down cow flow and add time to milking.

GOAL

- The parlour should be a place where cows and milkers are happy and content

FACTS

- Rubber matting on the parlour floor adds to cow comfort
- A large exit area at the front of the parlour will speed up cow flow
- Good light is important in the parlour

WATCH OUT!

- Narrow exits from the parlour slows down milking

Cows like rubber matting on the parlour floor

Clean udders and teats

Reducing risk of infections

It's important that udders and tails are kept trimmed and neat, especially for herds which milk through robotic milking systems. Robotic herds know that mastitis increases if tails and udders are not kept trimmed.

Hairy udders trap dirt which will increase the risk of mastitis occurring. This adds to the amount of work that milkers have to carry out during milking. Udders can be either clipped or singed.

Cows do not like the smell of burning hair and may be reluctant to come in to the milked. Make sure that only a yellow fame is used and that it is waved gently about six inches under the udder.

GOAL

- Trim and clean udders

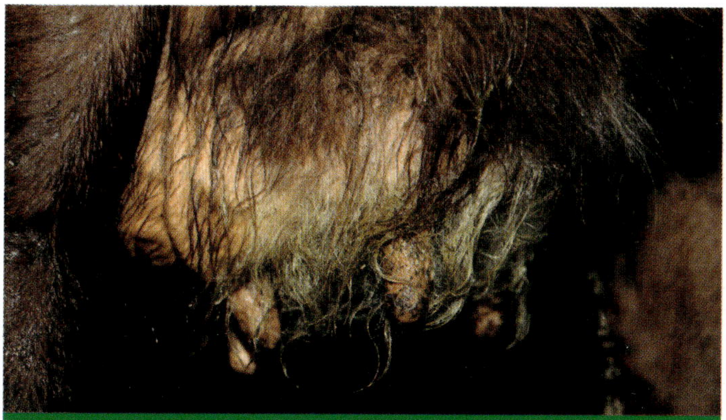

A hairy udder traps dirt and adds to the milking routine

An udder singer is an easy way to burn off hair

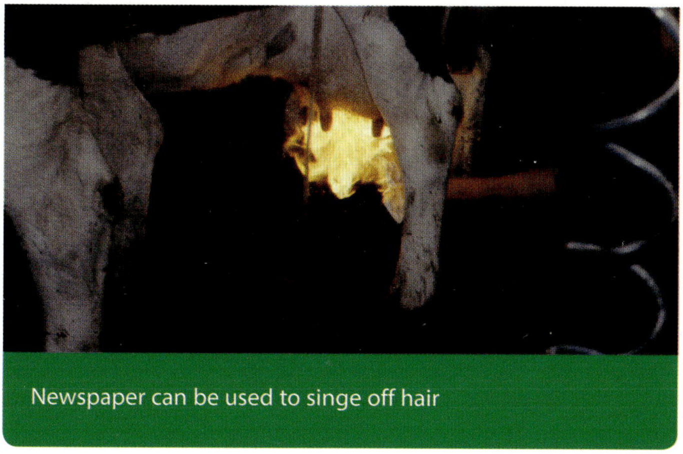

Newspaper can be used to singe off hair

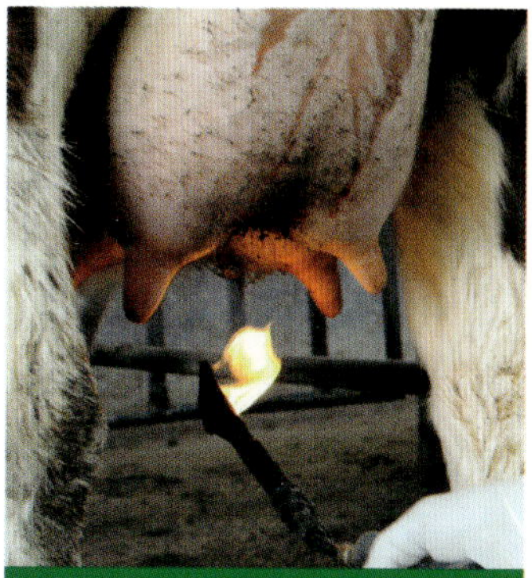

Always use a yellow flame to avoid burning teats or the udder

Clean udders and teats

Reducing risk of infections

Long tails lie in muck and then splash this dirt onto the rear teats and there will be an increased risk of mastitis. Long tails are unpleasant for the milkers, especially if they hit them in the face. Tails should be kept neat and trimmed.

Long tails get dirty and contaminate the udder and teats

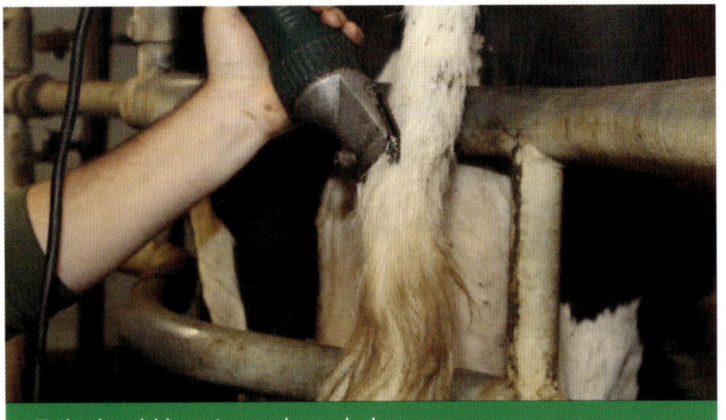

Tails should be trimmed regularly

Many farmers trim udders and tails if cows are separated for insemination or hoof trimming. This way it's easy to keep on top of the job.

FACT

- Long tails splash dirt onto the rear teats

TOP TIP

- Never miss an opportunity to trim udders and tails

Minimising the spread of infection

Gloves

All milkers should wear gloves that are kept clean throughout milking. This is because it is impossible to disinfect hands. We know this as surgeons scrub their hands with a disinfectant solution and then wear surgical gloves for operating.

UK research back in the late 1970s has shown that half of milkers hands were contaminated with mastitis organisms before milking, and at the end of milking all hands were infected.

Gloves are a hostile surface and if kept rinsed throughout milking will minimise the risk of hands spreading mastitis bacteria. If gloves become worn or torn they should be discarded. The wearing of gloves is a key control measure for contagious mastitis bacteria which cause high cell counts.

If gloves are not available, as is often the situation in some developing countries, it is essential to wash hands thoroughly before milking and dry hands using paper towels. While this will not stop the spread of infection, it will help to reduce this risk. In herds which hand milk, hands should be thoroughly washed between milking cows.

GOAL

- All milkers should wear gloves throughout milking

FACTS

- It is impossible to disinfect hands
- All milkers hands contain mastitis bacteria at the end of milking
- Gloves are a hostile surface for bacteria

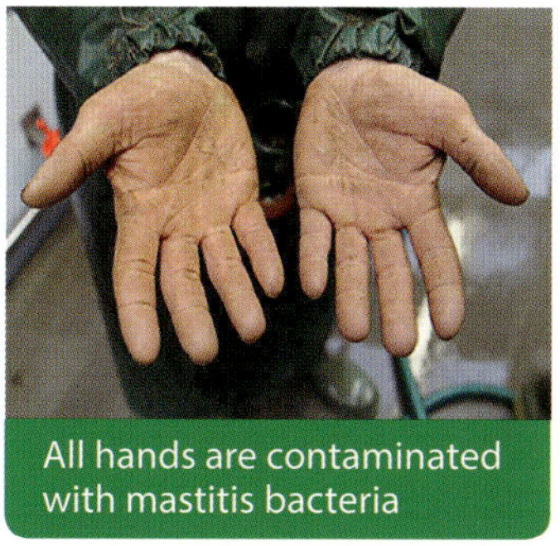

All hands are contaminated with mastitis bacteria

Alcohol does not kill all bacteria

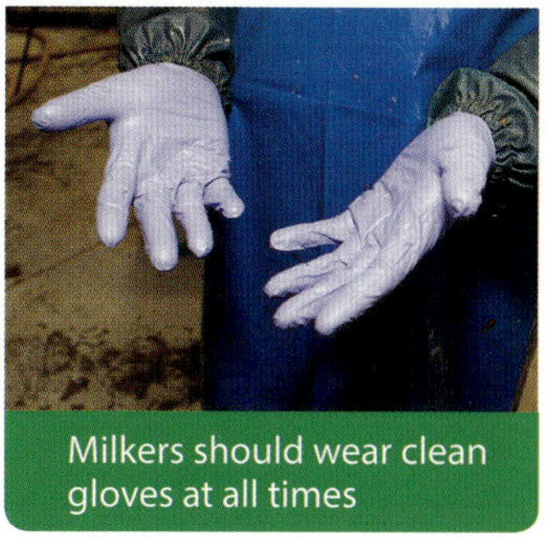

Milkers should wear clean gloves at all times

Gloved hands should be disinfected regularly

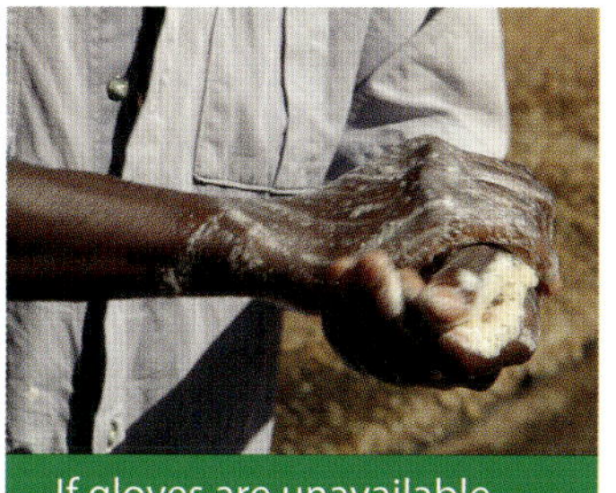

If gloves are unavailable, wash hands regularly

Dry hands thoroughly

Teat preparation

Aims of teat preparation

The key aim of teat preparation is to ensure that the teats are clean and dry before the milking unit is attached. This is to reduce the risk of clinical mastitis from environmental bacteria such as *E. coli*. Clean, dry teats will reduce the amount of environmental contamination of milk and helps to ensure that Bactoscan or TBC (Total Bacteria Count) remains low.

FACT

- Clean dry teats reduce mastitis

Cows should come into the parlour with clean udders and teats. If cows are kept in dirty conditions then the milkers job is very difficult and mastitis levels will be higher.

The problem here is the environment where the cows are kept

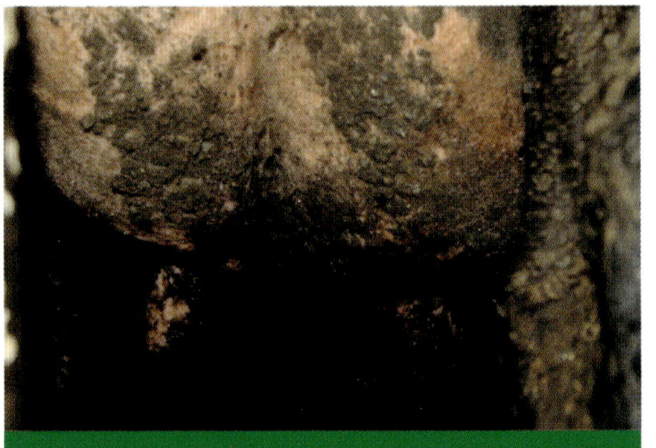

Teat preparation for a cow like this is challenging for any milker

Assessment of teat preparation in individual herds needs to be made in conjunction with the levels and types of clinical mastitis. For example if a herd has a low level of clinical mastitis then it would be unwise to make any changes to teat preparation, irrespective of how the teats are prepared, provided the cell count, Bactoscan or TBC are low. If a herd has a high level of environmental mastitis associated with poor teat preparation then changes must be made.

In some countries the milking unit is just attached to the teats without any preparation or handling whatsoever. This is often called 'dry cupping' and is commonly practised in New Zealand pasture based herds during the summer season. At this time cows teats are clean and levels of mastitis are low without any teat preparation.

In some herds the milking unit is attached without any preparation whatsoever

GOALS

- Clean dry teats before the milking unit is attached
- Cows should always have clean udders and teats

Teat preparation
Dry wipe

Dry wiping is where the teats are wiped with paper towel to remove any dirt or contamination. Best practice is to use one piece of paper towel per cow. This is often used in cows on sand cubicles (freestalls) where the teats are clean. The paper towels easily remove sand. Sand can become quite difficult to remove once you wet it.

GOAL

- With dry wiping best practice is one piece of paper towel per cow

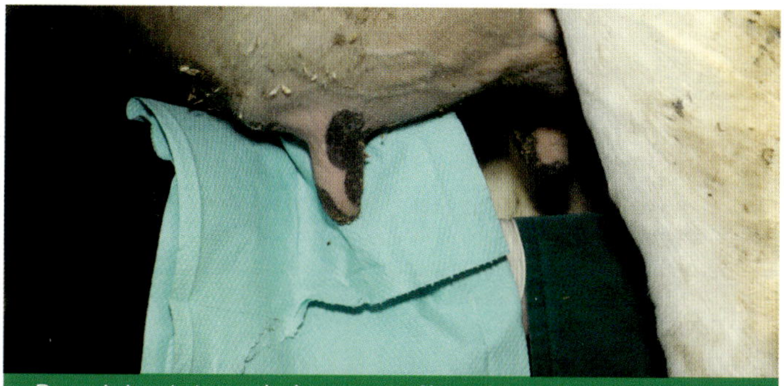

Dry wiping is intended to wipe off any loose dirt on the teats

Paper towel must be clean and dry for wiping teats

This paper towel could become wet and contaminated. Best to keep in a paper towel dispenser

Teat preparation

Washing teats

Teats can be washed with running water. This can prove effective in cleaning teats. The golden rule is that if you wash teats they must be dried. If you wash but don't dry them you mix the bacteria on the teat into a soup with the water. This then;

- Increases the risk of liner slip or impact forces which drive milk particles through the teat canal up into the udder
- Increases the risk of clinical mastitis
- Increases the amount of environmental bacteria entering the bulk tank
- Increases Bactoscan or TBC

TOP TIP

- If you wash teats they must be dried

FACTS

- Liner slip will increase on wet teats
- Clinical mastitis increases when teats are washed but not dried

If you wash teats with water they must be dried

This bucket of paper towels shows that the teats were dirty before milking

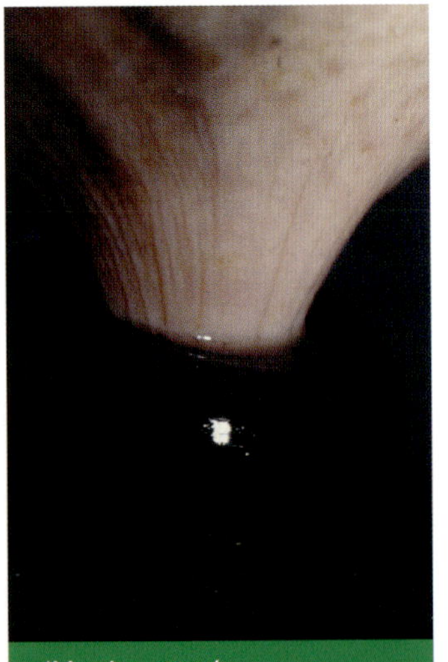

'Magic water'.
One minute it is there, the next minute it has been sucked into the liner

Teat preparation

Medicated towels

Teats can be cleaned using individual towels which are wet and impregnated with an alcohol solution. These wet towels help to remove any dirt, the alcohol helps to kill bacteria and they are easy to use. These towels can be quite costly.

Some farmers have tried to overcome these cost issues by making their own home made medicated wipes instead of using alcohol solutions. The author has dealt with an outbreak of clinical mastitis where the source of infection came from contaminated solutions used on home made medicated towels.

There is a risk of contamination and a rise in mastitis when home made medicated wipes are used

Damp cloths

The use of damp cloths soaked in a disinfectant solution or water is popular with some farmers to keep down the cost of teat preparation. Teats are wiped with a damp cloth which removes any dirt quite easily. One cloth is used per cow and these are then washed after every milking. One of the problems of using wet cloths is that while the teat might be clean after it is wiped, it is not dry and environmental bacteria might still be on the teats.

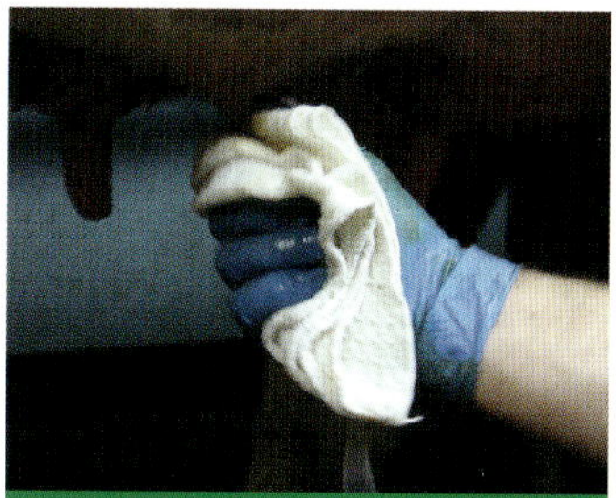

Wiping a teat with a damp towel

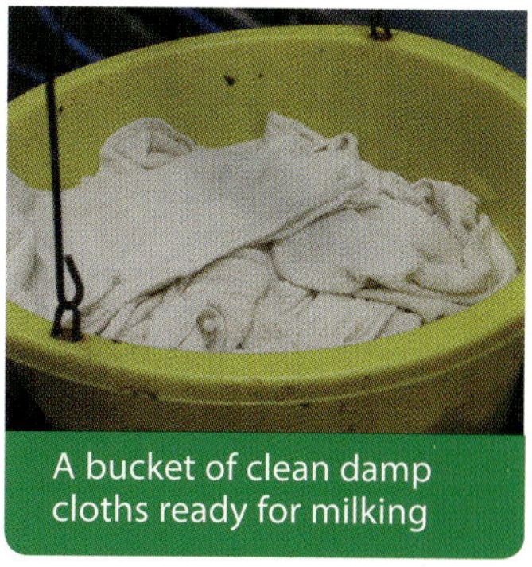

A bucket of clean damp cloths ready for milking

Common cloths which are used to prepare teats of several cows should **never** be used as they will spread infection from cow to cow. This will happen even they are soaked in a disinfectant solution. It takes time to kill bacteria.

Clean cloths after being washed. They will now be dried

Washing machines and tumble driers are needed when using cloths

Common cloths should NEVER be used as they spread infection from cow to cow

Teat preparation

Automated teat washers

There are some automated teat washers which have rotating brushes and spray water or a disinfectant solution to clean the teats. The brushes are pushed up each teat to remove any dirt. The water or disinfectant solution runs continuously through the brushes as the teats are cleaned. This cleans but does not dry the teats.

The equipment is expensive and can be cumbersome and heavy to use. Some brushes can be painful to cows. These wash but do not dry teats and so if used, teats should be dried. These types of devices are not recommended for use in milking parlours where alternative types of preparation can be used.

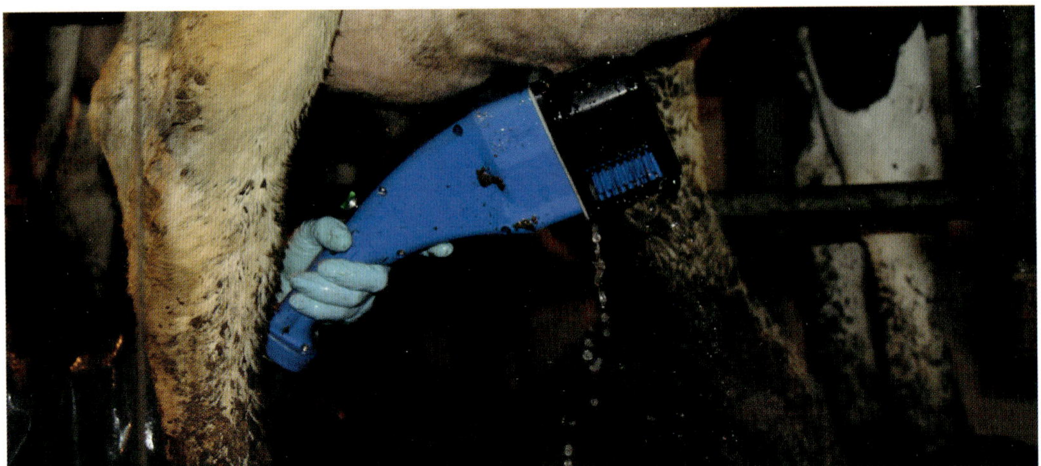

Automated teat washer using rotating brushes to clean teats

Brushes can be painful to teats and need to be kept clean throughout milking

Teat preparation

Predipping

Predipping is the gold standard for teat preparation and is where the teat is soaked in a disinfectant solution before milking. A contact time of at least 20 to 30 seconds is allowed. This contact time allows the solution to soften any dirt that is present and the chemicals in the predip help kill bacteria. The predip solution must be wiped off the teat using paper towel so you end up with a clean dry teat and so there us no chemical contamination of milk from predip solutions.

Predipping removes environmental contamination and reduces clinical mastitis due to *E. coli* and the other environmental bacteria. It also reduces the number of environmental bacteria entering the bulk tank protecting the Bactoscan or TBC.

There is plenty of research work to show that predipping is effective in reducing clinical mastitis. Herds which have poor teat preparation and start predipping see an immediate reduction in clinical mastitis. This is because environmental bacteria are removed from the teat before the milking unit is attached stopping these from entering the udder during the milking process. Some environmental bacteria can cause mastitis quickly after they enter the udder.

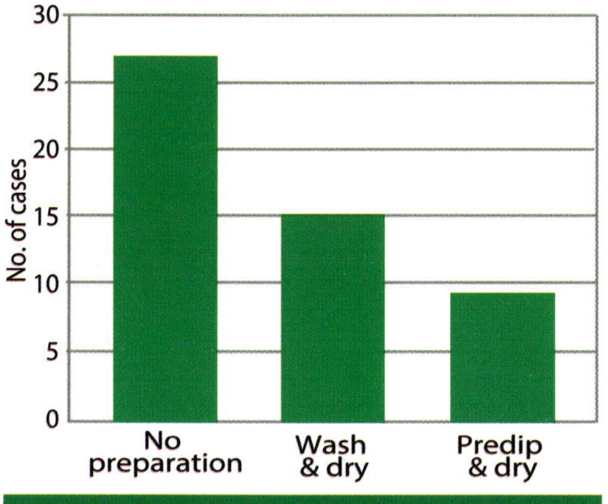

Effect of predipping in reducing the number of new clinical cases in four commercial dairy herds

This trial shows the number of cases of mastitis where there was no teat preparation, teats were washed and dried and predipped and dried. The cows that were predipped had the lowest levels of mastitis which was almost half that of cows that were washed and dried and a quarter than of those that had no preparation whatsoever.

Teat preparation

Predipping

Predipping is illegal in certain countries which do not allow disinfectant solutions to be applied to teats prior to milking, even if these are wiped off.

You need a disinfectant solution with a rapid speed of kill in a predip. Some disinfectants such as sodium hypochlorite are easily inactivated when they come into contact with environmental contamination. It is best to use a quality predip. Many predip solutions on the market do not have proven killing properties.

The predip should be applied to the entire surface of the teat. This means that it is best applied as a dip and not sprayed on. One of the problems with applying predip through a spray system is that predip can drain down from the base of the udder and onto the teat after it has been wiped dry. This means that chemical residues can enter the milk. If this solution is contaminated with mastitis bacteria because the base of the udder is dirty then you also increase the risk of mastitis.

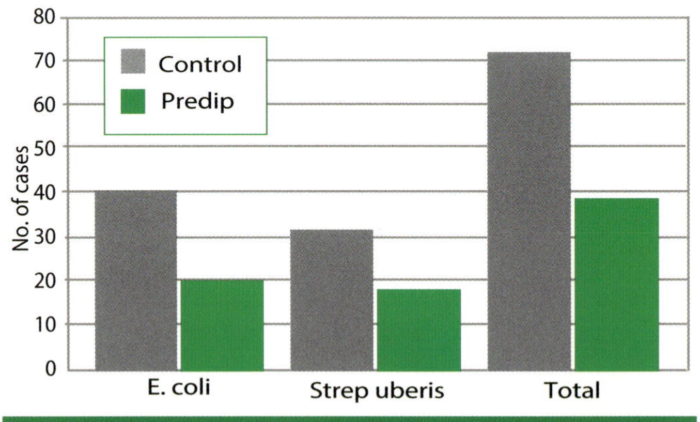

Effect of preddiping compared with conventional teat preparation on the incidence of clinical mastitis

This trial concluded that predipping reduced the amount of mastitis by over 40%.

TOP TIPS

- Allow a contact time of 20 to 30 seconds with predip then wipe dry
- Predip should be applied to the entire surface of the teat

FACTS

- Predipping is the gold standard for teat preparation
- Reduces clinical mastitis from environmental bacteria

Teat preparation
Predipping

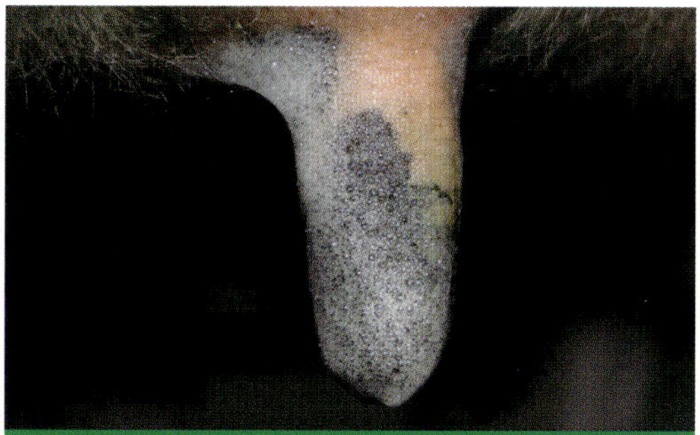

The entire surface of this teat has been covered with a foam predip

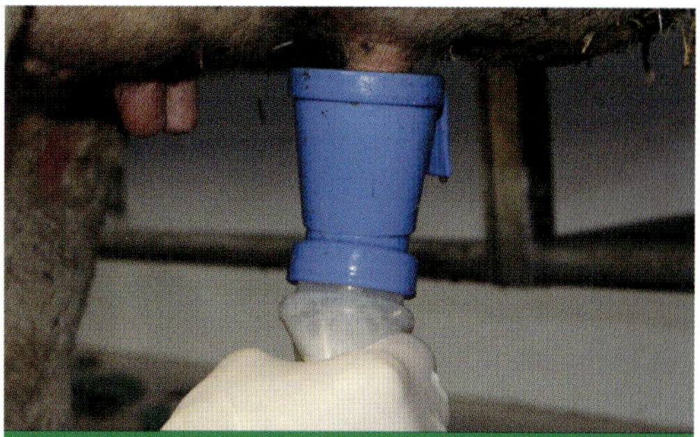

Predips are best applied using a teat dip cup

Teat dip cups should be cleaned after every milking and the old solution discarded

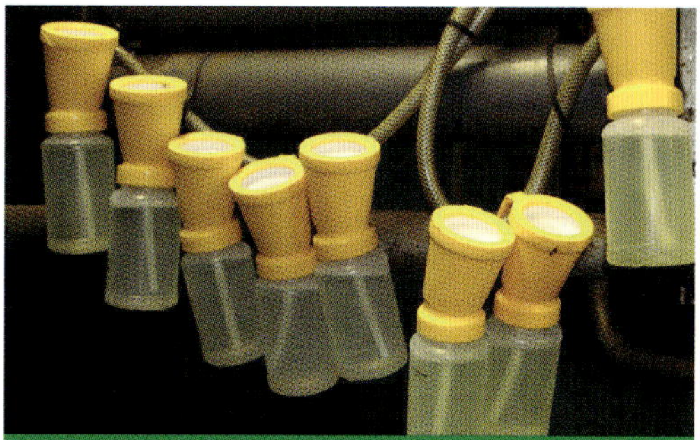

A row of teat dip cups full of predip ready for the next milking

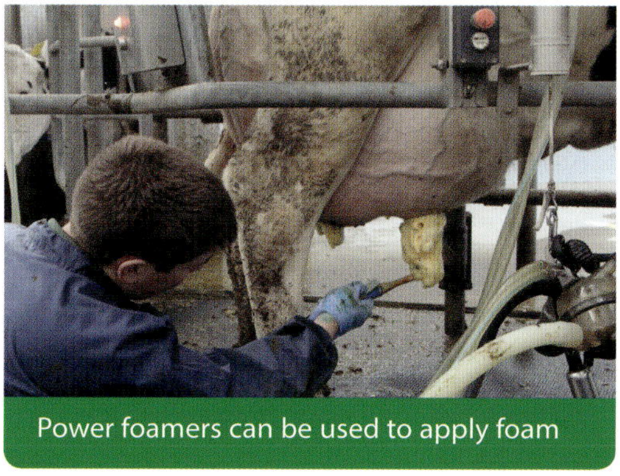

Power foamers can be used to apply foam

You can use liquid or foam solutions. Foam solutions are popular as they use less predip and they are good at wetting the teat. An added benefit from using foams is that the milker can easily see the solution on the teats and so the risk of teats not being dried should be minimal. These can be easily applied using power foamers or using a foaming teat dip cup which you squeeze to create the foam.

Drying teats

Individual towel per cow

Teats must always be clean and dry before the machine is attached to minimise mastitis and protect Bactoscan or TBC. You should use one piece of paper towel or a dry cloth per cow to dry teats. This is to ensure that you are not spreading infection from cow to cow. This is important in the control of cell counts.

Under no circumstances should paper or cloth towels be used on multiple cows. You can use good quality absorbent paper through towel dispensers so that they do not become contaminated. In developing countries old newspapers are commonly used for drying teats.

Milkers should be praised if they are carrying out good teat prep. We are all good at telling people if they are doing a bad job, and not so good at telling them that they are doing a good job. Milking cows is not an easy task and motivating milkers is really important.

If teat prep is poor, then it is important to take steps to ensure that this is corrected. Milkers should be reassessed shortly after changes are made to ensure that the routine has improved. This will require further training. Milkers should then be told that teat prep issues have been resolved and keep them motivated to ensure that it remains good.

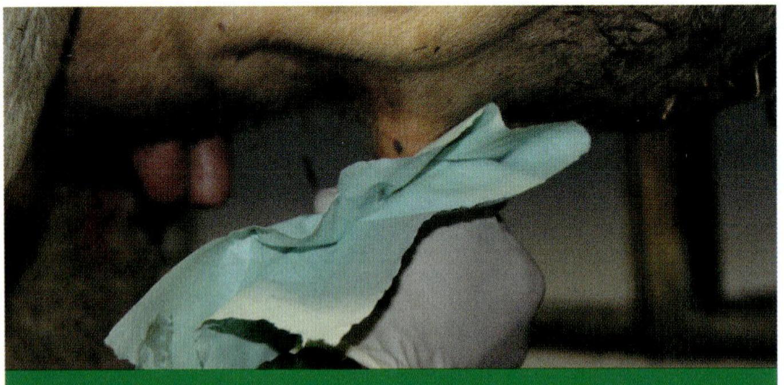

A new piece of paper towel should be used for each cow

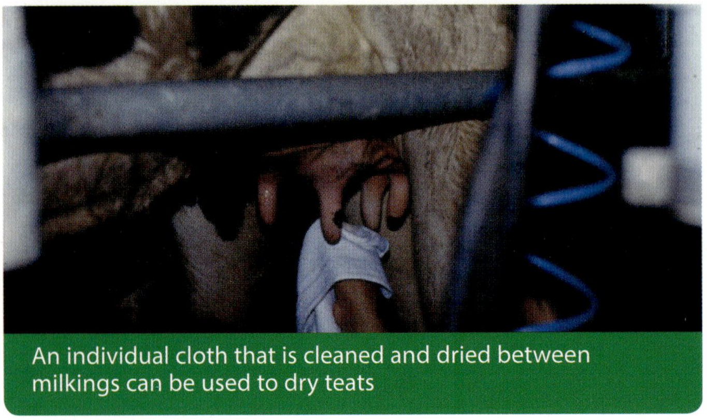

An individual cloth that is cleaned and dried between milkings can be used to dry teats

WATCH OUT!

- Don't use paper or cloth towels on multiple cows

TOP TIP

- It's important to assess teat preparation before the unit is attached

Assessing teat preparation

Milk filters

Check the milk filter or sock after milking. Many herds use disposable cloth filters which are discarded after every milking. In larger herds, the socks might be changed during milking. These socks can be held back and assessed to ensure that they are clean.

Some people will have different ideas on what is a clean or dirty sock! Milk socks will be clean if teats are cleaned and dry before milking

TOP TIP

- Check the milk filter or sock after every milking

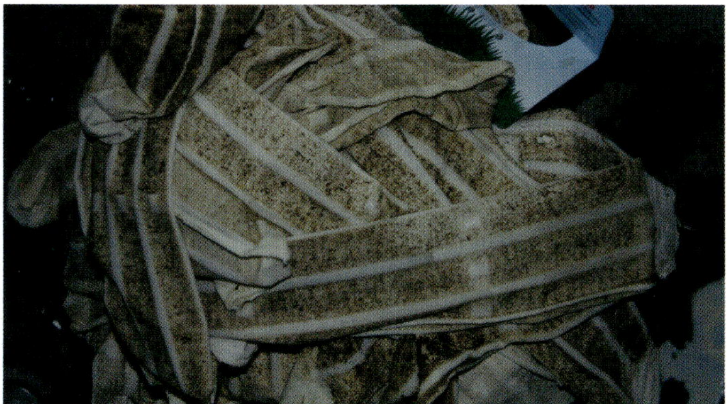

These milk filters have milked 100 cows and show significant amounts of dirt. These cows are in a very dirty environment

These milk filters are very clean having milked 200 cows

Some milk filters have a series of metal gauzes that are rinsed off after milking. In these cases, they need to be examined after milking and before they are rinsed off.

In some herds with multiple milkers, the milk socks or filters are assessed after each milking and scored. This helps to ensure that high standards of teat preparation can be maintained. This can be quite useful in assessing teat prep from previous milkings.

This is a reusable mesh milk filter

In large herds milk filters for different shifts can be kept and examined. These show the differences between milking teams

This re-usable milk filter was examined after milking and is spotless

Assessing teat preparation

Wiping teats after preparation

If you wipe teats with white paper towel immediately before unit attachment they should be clean.

One of the problems with this form of assessment in some herds is that if milkers know that this test is going to be carried out they might improve their teat preparation while you are present. They can then revert to their old practices when you have left.

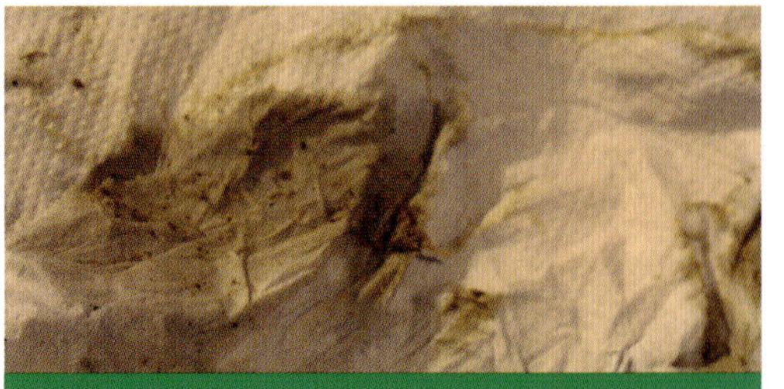

This piece of paper towel was used to wipe teats just before the unit was attached. See the amount of dirt left on the teats

Examine liners

Shine a pencil torch into the liners towards to middle to end of milking. There should be minimal amounts of dirt or faecal contamination present. Dirty liners indicate poor teat preparation.

Using a pencil torch to examine a liner during milking

Assessing teat preparation

Coliform count

The Coliform Count is an assessment of levels of coliforms in bulk milk which are an indicator of faecal contamination. Most coliforms come from dirty teats but some can come if the milking cluster falls into some manure and this gets sucked into the bulk milk

The target is to have less than 20/ml of milk. This is a laboratory test that is carried out on bulk milk samples that are kept refrigerated from bulk tank to the laboratory.

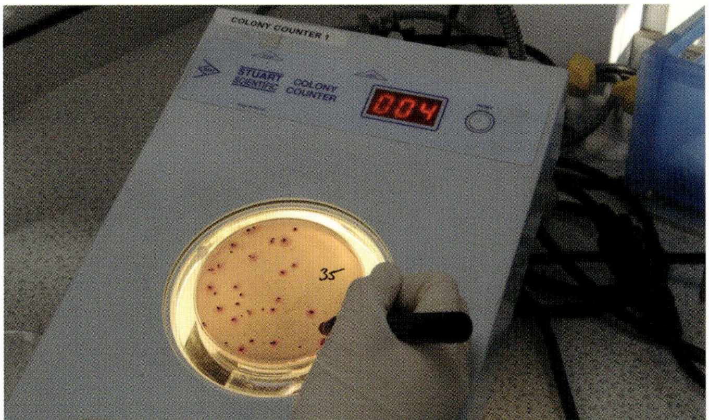

A coliform count gives a numerical indication of contamination of milk with coliforms

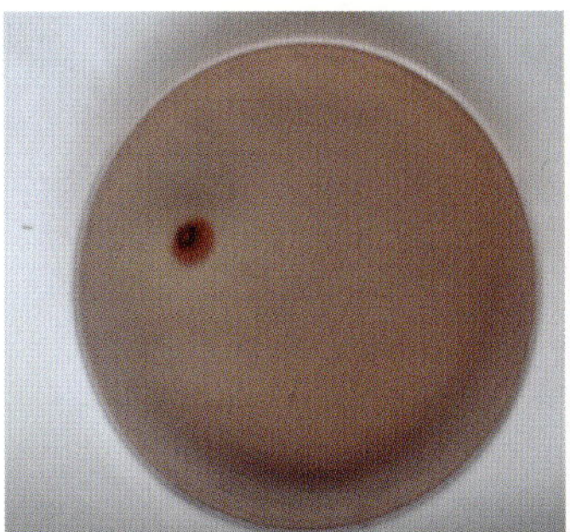

A coliform count of 1 indicates excellent teat preparation

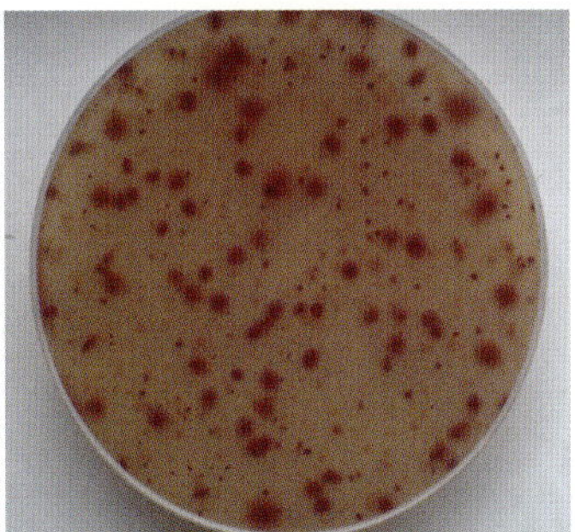

This coliform count is in the hundreds and indicates poor premilking teat preparation

One of the advantages is the coliform count is not subjective as it gives a numerical result. It can be used as a monitoring tool.

TARGET

- Coliform count of less than 20/ml

FACT

- A high coliform count will increase the risk of environmental mastitis

Mastitis detection

What is mastitis?

Mastitis means inflammation of the mammary gland. This results in changes to the composition of the milk such as clots, discoloured or watery milk. Sometimes the infected quarter will be swollen.

It's important to remember that clinical mastitis is the sign of the inflammatory response as a result of bacterial infection which entered the udder some time earlier. Also remember that in many cases, the cow's immune system will eliminate this infection without clinical mastitis occurring. Clinical mastitis occurs when the immune system is unable to eliminate infection.

It is important that mastitis is detected accurately and early to avoid the spread of infection to other cows. This can easily happen through contamination on the liners. Mastitis milk needs to be kept out of the bulk tank as it will be very high in somatic cells and bacteria and can spike the herd cell count or Bactoscan. The cell count of milk with clinical mastitis can be well over 10,000, compared to under 200 in a healthy quarter!

FACT

- The cell count of clinical mastitis milk can be over 10,000! A healthy cell count is under 200

How to detect mastitis

If in doubt about whether a cow has clinical mastitis or not, you can wait until the next milking to check the quarter, provided the cow is healthy and well. There is no need to carry out any other tests like the CMT (California Mastitis Test or Paddle Test) as this will only detect a rise in cell count which could be due to subclinical mastitis which is totally different to clinical mastitis.

Some farmers are very concerned that they must treat cases quickly to ensure that the cow will not die. There is no urgency for a number of reasons. Firstly, the majority cases of clinical mastitis are mild and secondly, the infection which causes clinical mastitis will have entered the udder most likely a couple of days before changes are seen in the milk. Accurate detection is important.

Confusion can occur if there are some clots or abnormalities in the foremilk, the first few strippings from each teat. If you get clots in the first squirt or two and then these disappear with further stripping, this is not clinical mastitis. Its a local reaction to infection in the milk in the teat canal and gland cistern - the place where some milk is stored in and above the teat.

If there are any clots or abnormalities in the foremilk, continue to strip this quarter. If there are still clots or abnormalities present after 8 or more squirts then this is mastitis as you are now stripping milk from the udder tissue. Most of the milk before this is from the teat and the gland cistern.

Cross section of the udder

Mastitis detection

How to detect mastitis

The most accurate way to detect clinical mastitis is by stripping the quarters and examining the milk. If the milk is normal then the cow does not have clinical mastitis.

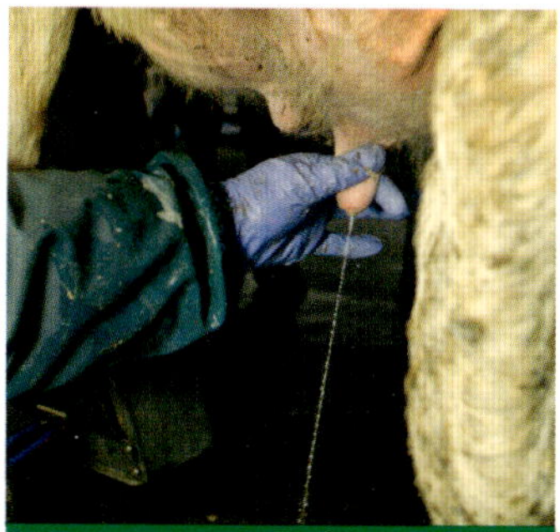

Stripping cows is the most accurate way to detect clinical mastitis

Large clots throughout the milk indicates clinical mastitis

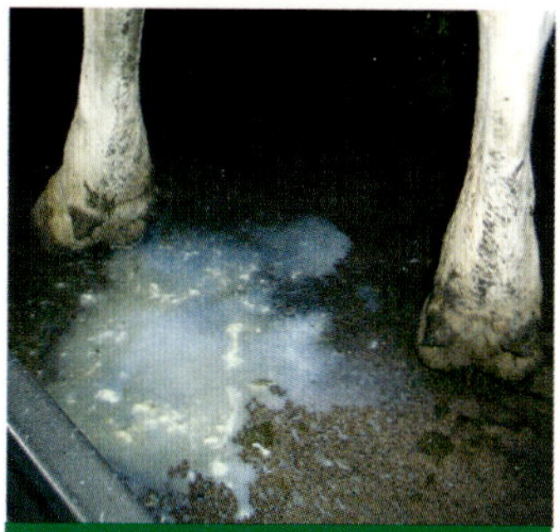

Smaller clots in the foremilk. Appearance of milk varies

E coli mastitis can be watery

TOP TIPS

- If clots in the foremilk strip clear this is not clinical mastitis
- Check at the next milking if in doubt whether a cow has clinical mastitis

FACTS

- Stripping quarters is the most accurate way to detect clinical mastitis
- A cow does not have clinical mastitis if her milk is normal
- You don't need to carry out tests to identify clinical mastitis

Mastitis detection

Mastitis detectors

Some people use mastitis detectors which are an in-line filter inserted in the long milk tube between the cluster and the milk transfer line or ACR flow meter. They are designed to pick up clots in milk.

There are a number of limitations of mastitis detectors:

- Not all milkers check detectors after each cow milked
- The milk has gone into the bulk tank before mastitis is identified and this can increase cell count and Bactoscan
- Not all milk with clinical mastitis produces clots, some cases produce a watery secretion
- Mastitis with very small clots might be washed through the filter
- Some milk will have bits of debris from the udder which could be mistaken for mastitis
- The filter can interfere with milk flow away from the cow and create vacuum instability

Mastitis detectors are considered an aid to mastitis detection. However, their negative points outweigh the positives and so their use in not recommended.

FACT

- Mastitis detectors are an aid to mastitis detection

A mastitis detector has a fine mesh filter to help pick up clots in milk

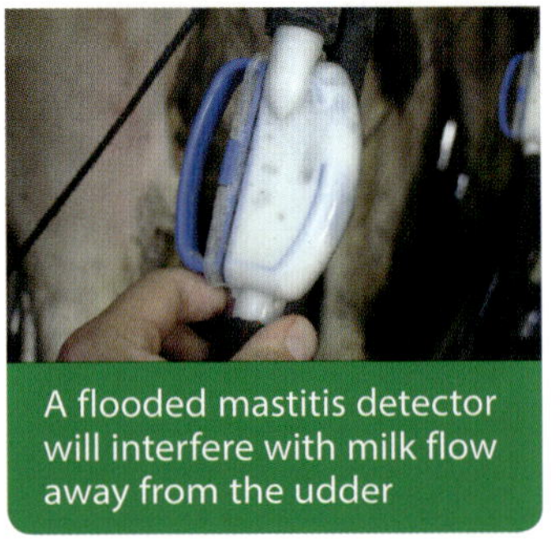

A flooded mastitis detector will interfere with milk flow away from the udder

Electrical conductivity meters

There are changes to electrical conductivity with mastitis but changes can also occur when cows are in oestrus or if they are ill. Electrical conductivity meters are fitted to some parlours as an aid to detection and check the milk flowing away from the cow.

Most robotic systems use individual quarter conductivity testing and compare this to the results from previous milking results and link this to changes in milk production and other parameters. This raises an alert so that the cow can be checked to see if she actually does have clinical mastitis.

Conductivity meters on their own are of little value for mastitis detection as they are not accurate enough.

FACT

- Electrical conductivity is not an accurate way to detect mastitis

Mastitis detection

Swollen quarters

A cow can come into the parlour with a swollen quarter but the milk appears perfectly normal. This can cause confusion. Does the cow have mastitis? Should she be treated?

If the cow is healthy and the milk is normal then no action should be taken. The swollen quarter is most likely due to an inflammatory response. Many of these cows come in at the next milking and the swelling has gone. This is because the inflammatory reaction has been successful in eliminating the infection. Nature has a great way of eliminating infections.

If the inflammatory response does not remove this infection then signs of clinical mastitis may develop and appropriate treatment or action can be taken.

FACTS

- You can get swellings in individual quarters as a result of injury or trauma

Diagnosis of gangrenous mastitis

Staph aureus and *Bacillus cereus* can cause gangrenous mastitis which kills the blood supply to the infected part of the udder. The affected quarter will feel very cold to the touch as blood is no longer circulating. In some cases you can see a line below which the udder looks dark or blue due to stagnant blood. Eventually, the affected udder tissue dies and falls off.

Many cases of gangrenous mastitis are not life threatening as it is more of a local reaction in the udder. Just like gangrene that occurs with mountain climbers resulting in loss of fingers or toes.

Just be very careful as you can also get changes to the colour of the udder with severe bruising and so you need to be able to distinguish between the two.

FACTS

- Gangrenous mastitis is cold to the touch
- Most cases of gangrenous mastitis are not life threatening

Foremilking

Stimulates the let-down reflex

Foremilking is where the milk is stripped from each teat before milking. It allows accurate mastitis detection and stimulates the milk let-down reflex. When the cow comes into the parlour she knows that she is going to be milked and there is a release of oxytocin, the hormone that causes milk let-down. This is called a conditioned reflex - it happens automatically. This conditioned reflex comes from stimulation of eyes, ears and what the cow thinks. You can increase the amount of oxytocin by intentionally stimulating the teats. This is called an unconditioned reflex and will maximise the let-down reflex.

FACTS

- You increase oxytocin levels by stimulating the teats
- It takes about a minute to get high oxytocin levels following stimulation

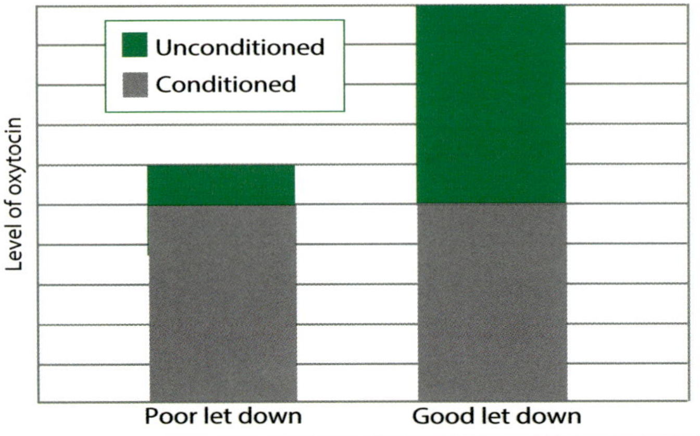

Effect of teat stimulation on oxytocin levels

The graph above shows that the greater the teat stimulation the higher the levels of oxytocin and the faster and more completely the cow will milks out. High levels of oxytocin are putting more pressure on the udder to force milk out through the teat.

It takes about a minute from teat stimulation to peak levels of oxytocin being released into the blood stream. It is best to attach the milking unit within 60 to 90 seconds of stripping the cow. This is important for two reasons: firstly to make the most of the high levels of oxytocin and secondly, the oxytocin action starts to decline and after about 12 minutes levels are low and milking time will be prolonged.

Let-down reflex

Speeding up milking

The effect of good teat stimulation can be clearly seen in two milk flow graphs from cows giving the same amount of milk. The first graph below shows milk flow rate from a cow with no teat stimulation. As soon as the cluster is attached there is a flow of milk, as shown by the green arrow.

After a minute, the milk flow almost stops, as shown by the red arrow. This is because as soon as the cluster is attached it removes the milk from the teat and gland cistern. This can be removed without the need of a let-down reflex. Milk flow starts after a minute as oxytocin levels started to rise. The cow milks out about 14 litres over a period of seven minutes

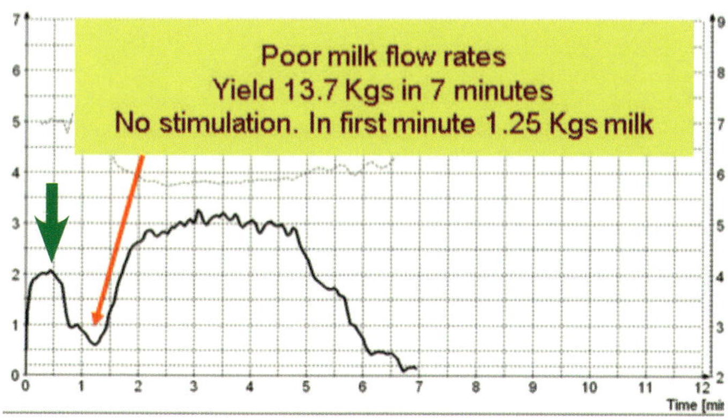

The graph below shows the milk flow rate for a cow with excellent teat stimulation where the milking unit has been attached within a minute. This cows milks out 14 litres in 4 minutes. Her peak flow rate is about 7 litres per minute compared to just under 3 litres per minute for the cow above with poor stimulation.

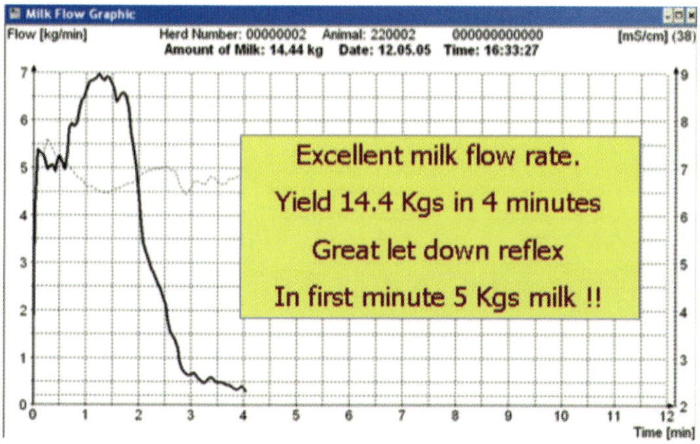

Teat stimulation

Speeding up milking

Foremilking can help speed up milking while allowing early and accurate mastitis detection. If a cow is predipped, it's recommended to apply the predip and then strip the teat as it's much easier stripping a wet than a dry teat. It's just more comfortable for the cow.

A summary of six separate trials have shown the benefits from good teat stimulation. These trials compared milking performance in cows with no stimulation before unit attachment to cows which had at least 20 seconds of stimulation and where the unit was attached within 60 seconds

	No stimulation	Optimal stimulation	Difference
Milk yield (litres)	10.4	10.8	+4%
Milking speed (litres/min)	1.8	2.1	+17%
Machine time on	6.3	5.5	+15%

This table shows that milk yield will rise by 4%, milk flow rates are 17% faster and it takes 48% less time to milkout the cow.

Unit attachment

Importance of timing

The cluster should be attached within 60 to 90 seconds of teat preparation. It should be attached without leaking in lots of air to ensure that vacuum levels remain stable. It is important to check the milking cluster is correctly aligned so that it sits evenly on the udder. If clusters twist there is the risk of teat slip or poor milkout as the liner can twist inside the cluster shell and partially close a teat.

TOP TIPS

- Attach the cluster within 60 to 90 seconds of teat preparation
- Ensure the cluster sits squarely on the udder

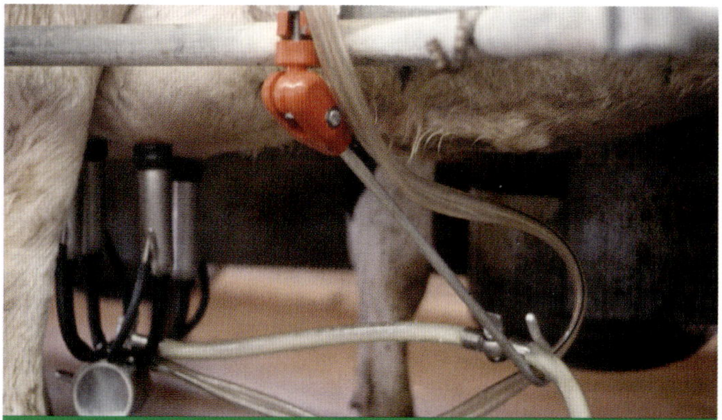

This parlour has a support arm to ensure good cluster alignment

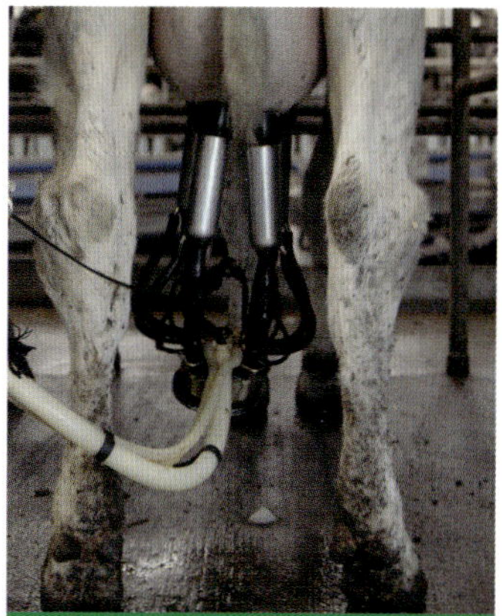

Good alignment in a rotary parlour

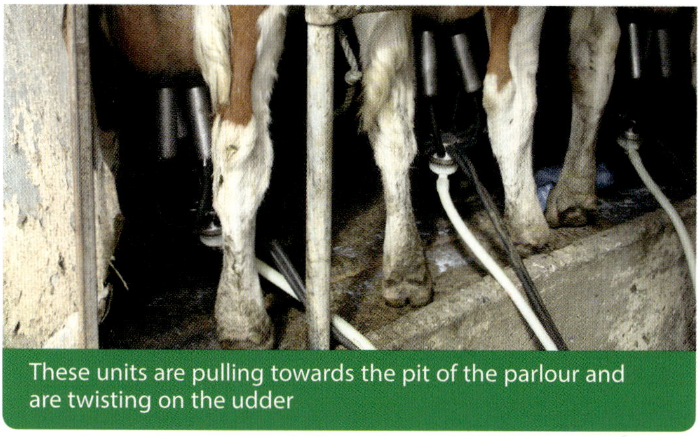

These units are pulling towards the pit of the parlour and are twisting on the udder

Batch preparation

Rotary and herringbone parlours

Cows should be prepared so that the cluster is applied to take advantage of the milk let-down reflex. In a rotary parlour this is very easy if there are two or more milkers dealing with teat preparation. The first can predip and strip: the second, who should be about a minute away from when the cow is prepped, and can dry and attach. This minimises the amount of time the cluster is on the cow and reduces the risk of teat damage and mastitis.

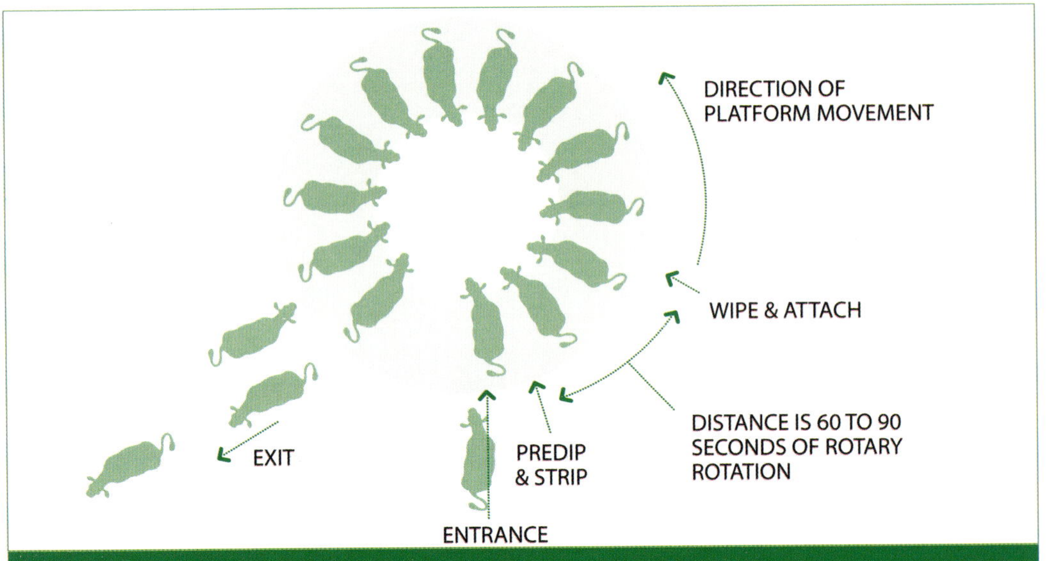

Rotary parlour

Cows might be prepared in small batches in herringbone parlours. So a 16 x 16 herringbone parlour, where there are 16 cows standings and 16 milking units, will have 8 cows standing on each side of the parlour. The row can be split into two groups of 4 cows which are prepared as a individual batch as shown in the diagram below.

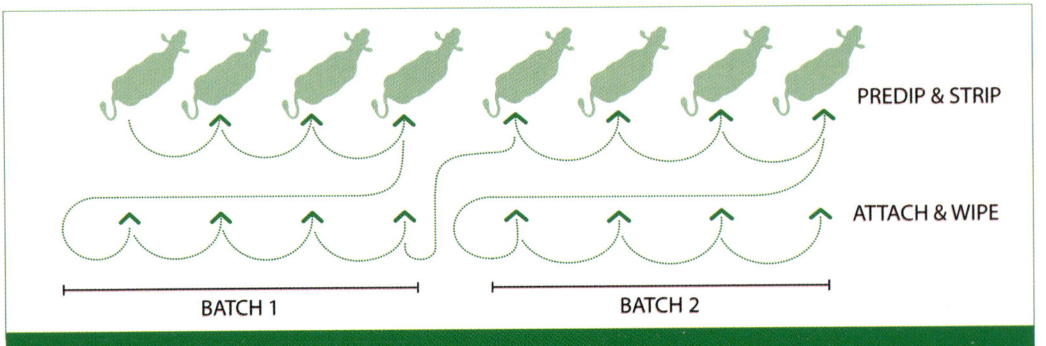

Herringbone parlour

In this routine the milker will predip and strip the first cow, then the second and repeat down the batch of four cows. He will then walk back to the first cow, dry the teats and attach cluster, move to the second cow and repeat down the line. He then moves on to the second batch of cows.

Batch preparation
Follower; leader system

A follower; leader system can be used in herringbone parlours with two or more milkers as shown below. The first milker predips and strips the first cow, then the second and moves down the line. About a minute later the second milker will dry the teats of the first cow and attach the cluster, repeat on the second cow and move down the line. The aim here is that the milking cluster is attached within 60 to 90 seconds of stimulation.

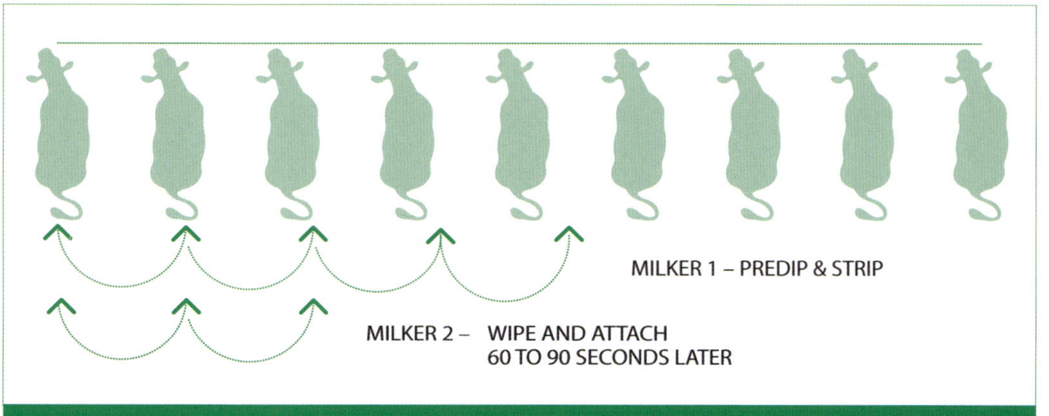

Herringbone parlour utilising a follower; leader system

It is not uncommon for milkers to walk miles in the parlour but not work efficiently. A set efficient routine helps to speed up milking.

Unit removal after milking

Avoid overmilking

The milking unit should be removed when the cow is milked out. The majority of modern parlours are fitted with automatic cluster removers commonly called ACRs. These measure milk flow rate and shut off the vacuum and remove the cluster at pre-set flow rates and timings.

The current recommendation from many manufacturers is that ACRs should be set to come off when the milk flow drops to between 400 and 500 ccs/minute in herds that milk twice a day (2x). For herds that milk three times a day (3x) the settings are increased to 600-800 ccs/minute. There are many newer parlours where units are set to come off at higher flow rates with no adverse effects on mastitis or cell counts.

Most new parlours allow adjustment of ACRs so that different milk flow rates and timings can be used to increase or decrease the flow rates and timings when the ACRs trigger the unit to be removed. Many of the older ACRs cannot be adjusted.

In herds without any ACRs, or where they are milking using old ACRs which are often set to low flow rates, the milkers should to watch each milking unit and remove these when the cows are milked out.

It's important to ensure that cows are not overmilked or the full vacuum level of the milking system will pull on the teat. This can cause hyperkeratosis, where excess keratin is produced in the teat canal and is then pulled out through the base of the teat. There is a greater risk of clinical and subclinical mastitis with hyperkeratosis.

Overmilking. There is no milk in the cluster at the end of milking.

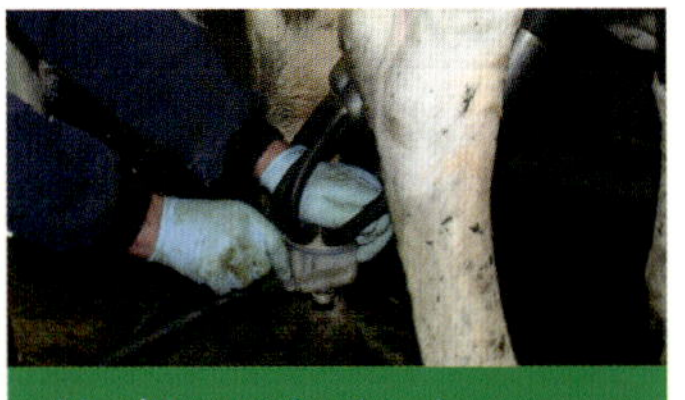

Machine stripping is unnecessary

In old parlours, machine stripping was commonly carried out at the end of milking. This is where the milker applied pressure to the cluster and massaged each quarter. This is totally unnecessary and will make no difference to milk production, but will increase the risk of teat damage and mastitis.

Units should never be pulled off without first shutting off the vacuum supply. Removing milking units that are still under vacuum will increase teat end damage and the risk of mastitis as you can create impact forces which can drive milk droplets and bacteria into the udder.

If the milking cluster gets dirty it must be hosed off with clean water so that the outside and the liners are clean before it is attached to the next cow. Cows can kick off units, or they can fall off and land in manure. If these are not cleaned before the cluster is attached to the next cow, the risk of clinical mastitis will increase.

FACT

- The milking cluster should be removed when the cow is milked out

TOP TIP

- ACR should come off when the flow rate drops to 400 and 500/mls per minute with 2 x milking and 600 to 800/mls with 3x milking

WATCH OUT!

- Machine stripping or pulling the unit off under vacuum is not recommended

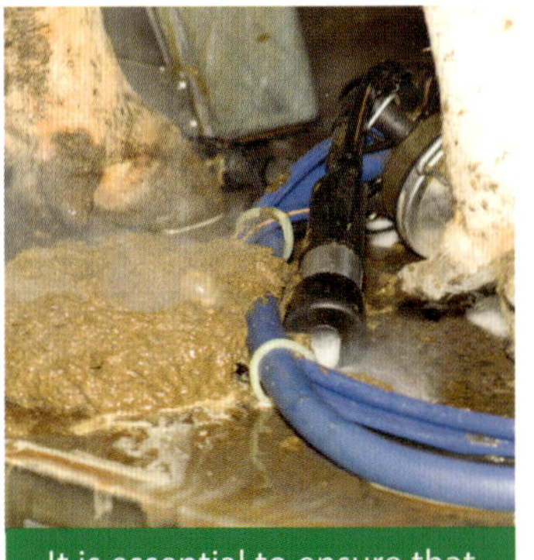

It is essential to ensure that clusters and their liners are clean before being attached to the next cow

Post milking teat disinfection

Introduction

Post milking teat disinfection is commonly called teat dipping or teat spraying depending on how the solution is applied. It has two aims; firstly to kill bacteria remaining on the teats after milking and secondly to help ensure that teat skin remains in optimal condition.

No matter how good your hygiene practices are during milking, there will always be spread of infection from cow to cow. If you apply teat dip solution then you reduce the new infection rate by 50%. The reason that this is not higher is that it is impossible to kill all bacteria on teat skin, just like you can't disinfect hands.

The herd in the diagram below started to teat dip half way through lactation. The reduction in the rise of subclinical infection can be seen in the red line. Teat dipping does not stop all spread of infection. Teat disinfection should take place after every milking throughout lactation as infection is spread every time that a cow is milked.

TOP TIPS

- Post milking teat disinfection is a key part of mastitis control
- Apply post-dip solution to every cow after every milking throughout lactation
- Do not compromise on the quality of a post milking teat disinfectant solution

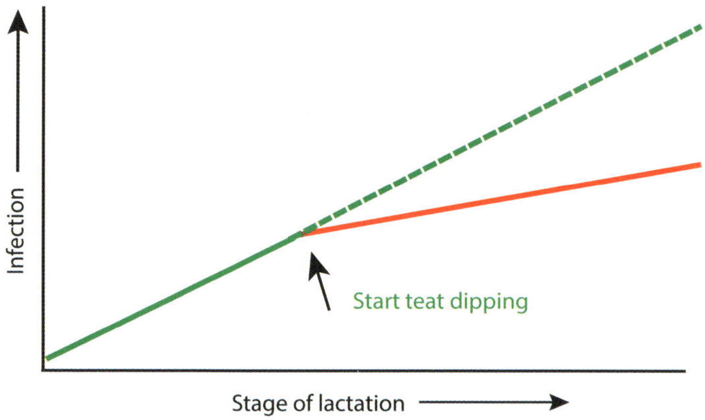

Manual teat dipping can result in excellent teat dip cover

Teat dipping is aimed at controlling the spread of contagious mastitis bacteria. It is essential that the entire surface of the teat is coated in teat dip solution as the liner has come into contact with the whole teat.

Post milking teat disinfection is an essential part of mastitis control.

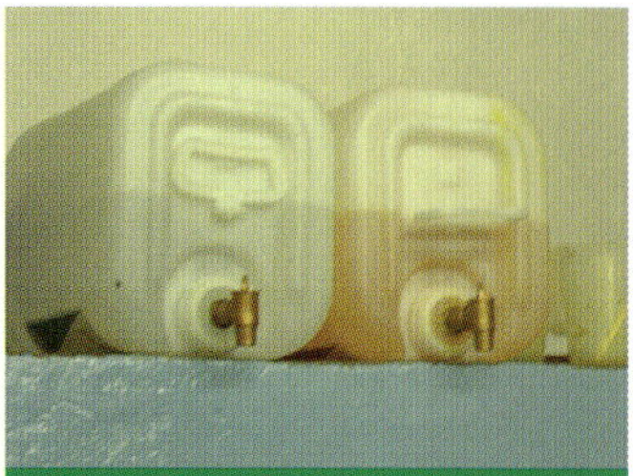

Teat dip solution should be stored correctly to avoid contamination

WATCH OUT

- Some of the cheaper teat dip solutions have poor teat skin conditioning properties

FACT

- Teat dipping reduces the new infection rate by around 50%

Post milking teat disinfection is covered in great detail in 'How to control Somatic cell counts', the first handbook in this series.

Cluster disinfection

Disinfection and back flushing

Cluster disinfection can be used after milking cows with high cell counts, clinical mastitis or after every cow is milked. The aim is to reduce the risk of spreading mastitis bacteria from cow to cow by reducing the number of bacteria on the liner. The majority of dairy herds with low cell counts do not cluster dip as they have subclinical mastitis under control.

Backflush systems will have little benefit in herds with clean cows and a good milking routine. However, they can have positive benefits in herds with poor hygiene and poor mastitis management.

To cluster dip efficiently, two shells should be immersed in the disinfectant at one time, removed and then the other two shells immersed. This is to avoid air becoming trapped in the cluster and stopping the solution going up into the liners. This can happen if the entire cluster is dipped into a disinfectant solution at one time. In the photos below the milker is cluster dipping correctly putting two shells into the bucket of disinfectant solution, and then the other two. This avoids air being trapped and allows the solution to come into contact with the entire liner surface.

Always keep buckets full of clean disinfectant available and dip the clusters after every cow is milked. The buckets and milking equipment are spotless.

Step 1 - 2 shells into bucket

Step 2 - add the other 2 shells

Full details about backflushing and cluster dipping after milking can be found within 'How to control Somatic cell counts', the first handbook in this series.

Teat canal closure

Encouraging cows to remain standing after milking

People were told they had to keep cows standing for 30 minutes after milking to allow the teat canal to close. In some cases, this meant that all cows were forced to remain standing for 30 minutes after the last cow was milked. So if milking took two hours, the first cow would have to remain standing for almost two and a half hours morning and afternoon! This will have a major impact on lameness and welfare.

By the time these cows have walked to pasture their teat canals will have closed

When cows are at pasture, they should exit the parlour and walk to the field where they will immediately start to eat. Feed intake is highest immediately after milking. This means that these cows will be standing for well over 30 minutes after the end of milking allowing plenty of time for the teat canal to close.

These cows have been kept standing until all the cows were milked

When cows are housed, they should leave the parlour and have immediate access to feed. Once these cows have eaten their fill they can go and lie down. The teat canal in the majority of cows will have closed at this stage.

If a cow is lame or not feeling well then she should be allowed to lie down immediately after milking even if there is a risk of mastitis occurring. These cows are going to be tired or have sore feet and so this must be allowed on welfare grounds. The alternative is that some of these cows might end up lying down in a dirty concrete yard where the mastitis risk will be significantly greater.

TOP TIPS

- Encourage cows to stand after milking by offering fresh feed
- Don't force cows to stand for excessive periods of time after milking

If cows are offered fresh feed after milking they will stand and eat giving time for the teat canal to close

Milking order

Large and small herds

In large herds there might be several groups of cows. The aim is to milk the most vulnerable cows first and then the problem mastitis cows at the end. This will minimise the risk of spreading infection from cow to cow and this is very important when dealing with herds with high cell counts.

Any cow under treatment or that has medicine residues where her milk needs to be discarded should always be milked last. This helps to safeguard accidental transfer of milk with residues and avoid contamination of the bulk tank.

Large herd

1. Freshly calved
2. Heifers
3. High yielding
4. Low yielding
5. High cell count (if the herd has a high SCC)
6. Cows under treatment

Small herd

1. Main group
2. High cell count (if the herd has a high SCC)
3. Cows under treatment

Milking routine SOP

Summary

It's important to have a written milking procedure and that all milkers understand what is expected and follow this. Cows like a consistent parlour routine and if each milker has a different way of milking this causes confusion. An example of a milking routine is shown below.

A good routine will ensure that cows are milked swiftly with minimal effect on mastitis and milk quality. It should be a really pleasant experience for cows and milkers alike.

TOP TIP

- Have a written SOP for the milking routine to ensure a uniform milking routine

SUMMARY OF THE MILKING ROUTINE

Goal - milk clean dry teats with minimal risk to udder health

- Wear gloves and keep them clean throughout milking. Change as necessary
- Batch prep your cows in batches of 5 to 8 so that the unit is attached within 60 to 90 seconds after teat preparation to maximise speed of milking
- Predip and strip the first cow, then the second and repeat down the batch of cows
- Go back to the first cow and wipe and attach the cluster and repeat down the batch of cows
- Ensure that the cluster is sitting squarely and evenly on the udder
- Move on to the next batch of cows
- Remove the unit to avoid overmilking. ACRs should be set to come off at a flow rate of 400 – 500 mls/minute for 2X milking and 600 – 800 mls/minute for 3X milking
- Teat dip the entire surface of each teat after every milking to maximise killing of bacteria that have been transferred during milking
- If a cow has mastitis, milk last or through a separate cluster that is disinfected after milking every mastitis cow
- Allow cow to exit from the parlour and encourage standing for 30 minutes by offering fresh feed after milking

Dry Period Infections

How to control clinical mastitis

7

Dry period infections

Are teats always sealed throughout the dry period?

For many years people thought that when cows were dried off the teat canal was sealed throughout the dry period. We now know that this is not the case. There is lots of trial work that shows that many teats remain open throughout the dry period.

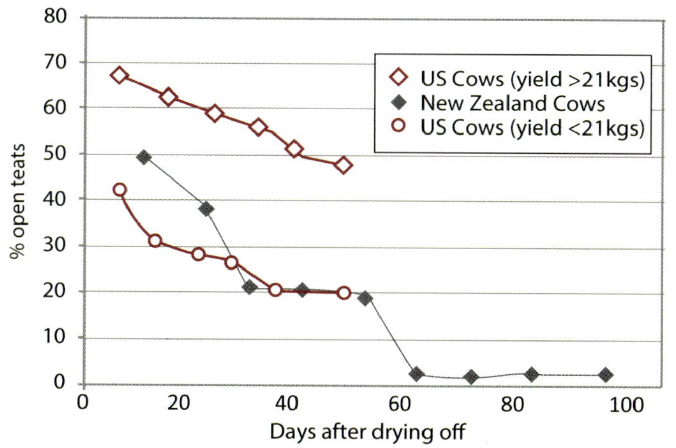

Trials to investigate percentage of open teats during dry period

FACTS

- At least 50% of teats are still open 10 days after dry off
- Half of teats from high yielding cows are open six weeks after dry off
- Bacteria can enter the udder during the dry period
- Always ensure the udders and teats of your dry and calving cows are clean to minimise dry period infections

The graph above shows that 50% of teats in New Zealand dairy cows were open 10 days after dry off. 20% of teats were still open 50 days after dry off. This study found that 97% of clinical mastitis cases that occurred in dry period were in quarters that had open teats.

US trial work shows that almost 70% of cows dried off giving more than 21 litres per day had open teats 7 days after dry off. 48% or almost half of these teats were still open at 6 weeks after dry off!

Higher yielding cows have faster milk flow rates and will have a higher risk of having open teats during the dry period. New Zealand cows are not the the highest yielders so this is why the US study showed more open teats.

Dry period infections

UK trial work

UK trial work was carried out on 629 dairy cows to assess the effect of whether coliform mastitis bacteria entered the udder during the dry period and their significance on clinical mastitis during the next lactation. All cows in this trial were dried off with antibiotic dry cow therapy with four different antibiotic dry cow products.

Sterile milk samples were collected from all trial cows at dry off. Samples were also collected from half of all the quarters at two weeks and one week before calving, at calving and whenever there was a case of clinical mastitis. Only half of quarters were sampled after dry off so that there were control quarters which are needed for comparison purposes.

The sampling procedure is shown below.

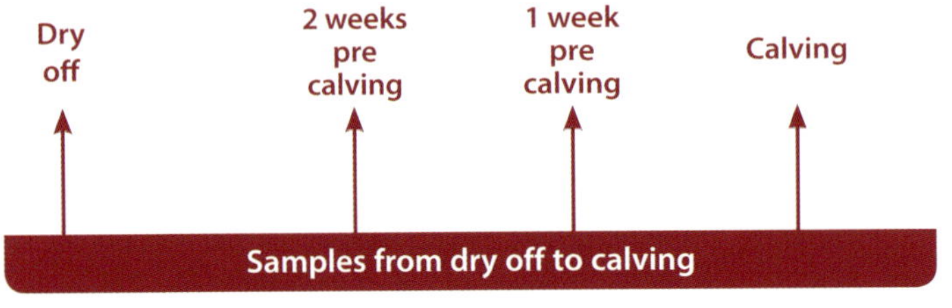

DNA fingerprinting was carried out on samples that cultured positive for *E coli*. DNA finger printing is like looking at a car number plate. Random numbers have been used to represent the *E. coli* DNA results.

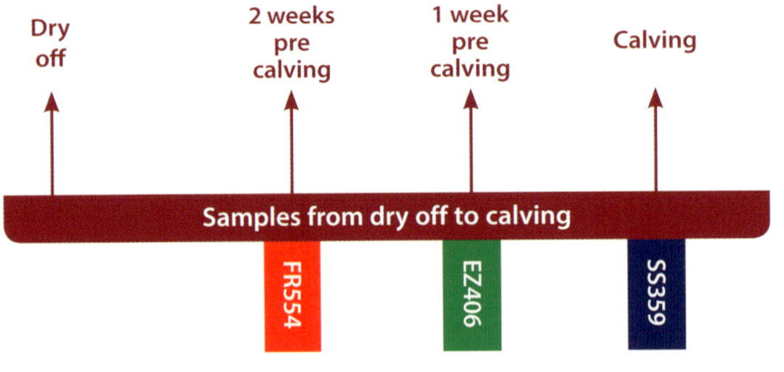

Dry period infections

UK trial work - results

All cases of mastitis were sampled and DNA fingerprinted. The data was checked to see if there was any association between the DNA strains identified. The data from the cow below shows that there is no relationship between any of the *E. coli* strains that caused clinical mastitis in lactation and those which entered the udder in the dry period.

FACT

- Dry period infections cause mastitis in the following lactation

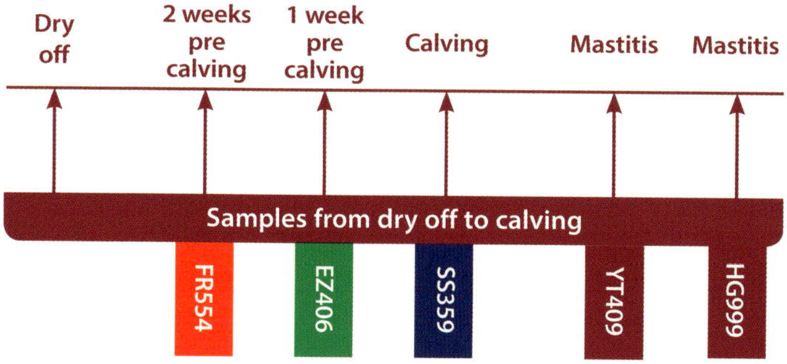

DNA fingerprinting from the cow below shows that strain FR554 was the cause of clinical mastitis in lactation and was picked up in the dry period. This shows that dry period infections cause mastitis in the following lactation.

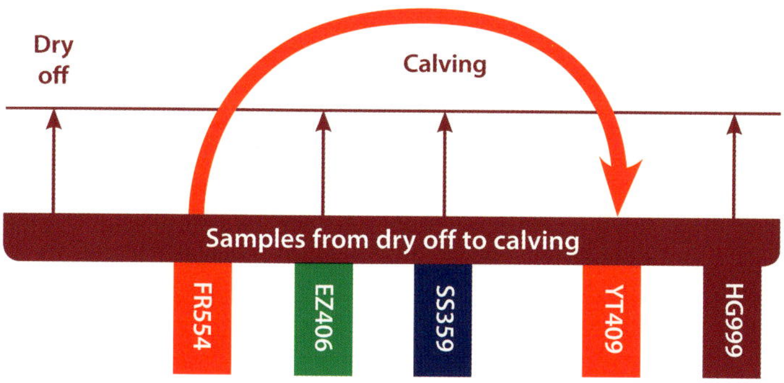

Dry period infections

Lactoferrin levels

The key question is why do these cows not get clinical mastitis in the dry period at the time that bacteria enter the udder? The explanation is simple. Levels of the enzyme lactoferrin rises in the dry udder. Lactoferrin binds up iron so that it is unavailable when the cow is dry. Iron is needed for bacterial growth and so this stops the bacteria from growing and causing mastitis. The bacteria just sit there in the udder.

Iron levels rise once the cow starts to start to bag up and come into lactation. This means that these bacteria can start to multiply; some might cause mastitis around the time of calving, others may remain dormant in the udder and do not cause mastitis for several months, while others will be eliminated by the immune system.

It's like going into a bar at 3 o'clock in the morning. You are in the right place for a drink but at the wrong time as the barman has gone home. This is what happens with dry period infections. The bacteria just enter and remain in the udder and they can't grow. The barman then comes to work and maybe you choose to have a drink then, or later or maybe change your mind and choose not to have a drink at all and leave.

This UK trial showed that 52% of all coliform clinical mastitis cases in the first 100 days of lactation originated from dry period infections. This work has been duplicated by many other research workers with not just coliform but also Strep uberis mastitis.

A summary of these key trial results shows;

- 52% of clinical coliform infections in the first 100 days originated from a dry period infection (UK work)
- 61% of new gram negative infections (coliforms etc.) entered the udder when the cow was dry (US work)
- 51% of all new environmental *Streps* entered the udder when the cow was dry (US work)
- 56% of clinical cases due to *Strep uberis* occurred originated from dry period infections
- 33% of *Strep dysgalactiae* clinical mastitis cases originated from dry period infections (UK work)

Dry period infections
Timing of new infections

The graph below shows that there is a significant increase in the number of new intramammary infections in the first and last two weeks of the dry period when the teat canal keratin plug is being formed and then when it is dissolving. There is a very low level of new infections in the middle of the dry period. The highest level of new infections occurs around the time of calving.

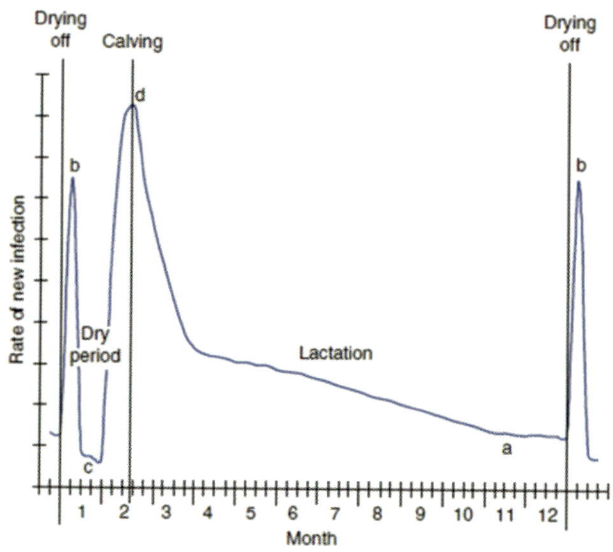

Timing of new infections over lactation and the dry period

The above graph shows the incidence of new intramammary infections in lactation and the dry period. This does not mean that this is when clinical mastitis occurs as many bacteria are eliminated by the cow's immune system and do not cause clinical mastitis. The high risk periods are in early lactation and the beginning and end of the dry period.

Dry period infections

Bacteria entering the udder during the dry period

The graph below shows a breakdown of cases of mastitis by month of lactation and whether they originated from the dry period or were picked up during lactation. The dark red bars represent mastitis from dry period infections and the light red bars represent clinical mastitis originating from lactation infections.

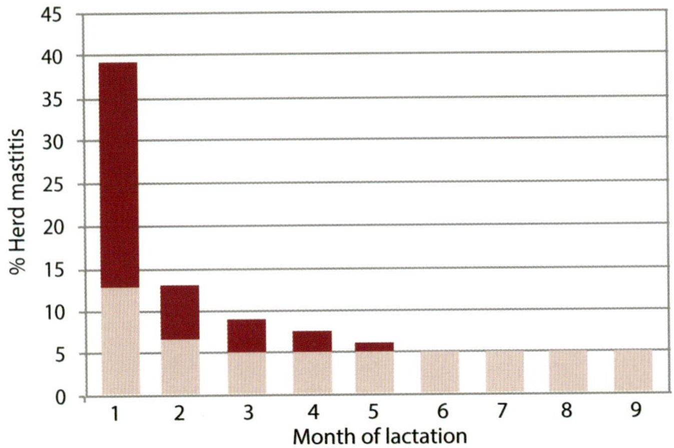

Breakdown of cases of mastitis by month of lactation

This shows that two thirds of clinical cases in the first month of lactation originated from dry period infections.

The chart clearly shows that dry period infections are the most significant cause of clinical mastitis in the first three months of lactation.

The need to seal the teat with an internal teat sealant, commonly called teat seals, throughout the dry period becomes very clear when you put all the above information together. Lots of teats remain open throughout the dry period. This allows bacteria to enter the udder during the dry period and cause clinical mastitis in the following lactation.

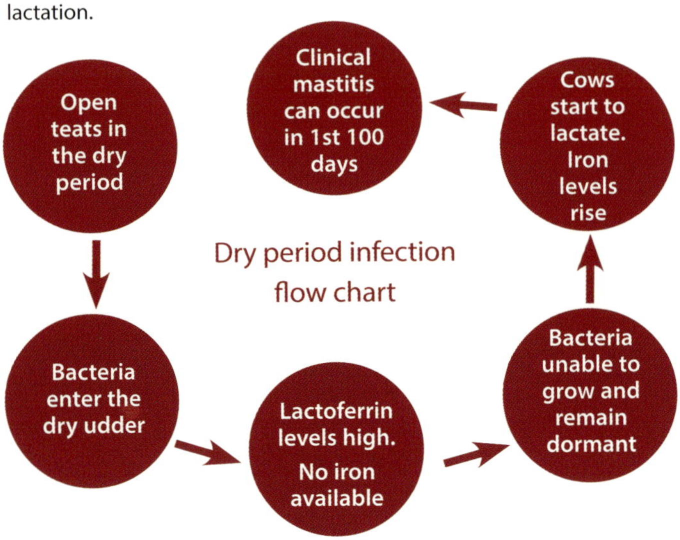

Internal teat seals

What is a teat seal?

Most teat seals contain bismuth subnitrite in a paste. The paste has no antibacterial or antibiotic properties and so has no effect on reducing subclinical mastitis or cell counts. A teat seal forms an artificial barrier at the base of the teat at dry off. It is infused into the udder at dry off and should be infused after antibiotic dry cow therapy, if this is used

Teat seals help prevent environmental bacteria entering the udder during the dry period. This will reduce the new infection rate and clinical mastitis in the next lactation. Teat seals will not seal all teats or prevent all bacteria from entering in the dry period.

FACTS

- Teat seals do not contain any antibiotics
- Will not reduce cell counts
- Reduce dry period infections
- Reduce clinical mastitis in the following lactation
- Will not seal *all* teats or prevent *all* bacteria from entering in the dry period
- Seals remain in the teat for up to 100 days

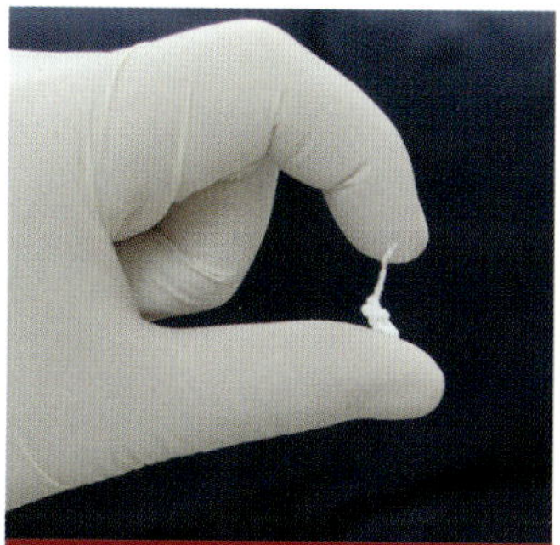

This shows the tackiness of an internal teat seal

This shows a cross section of a teat with an internal teat seal forming a plug in the teat

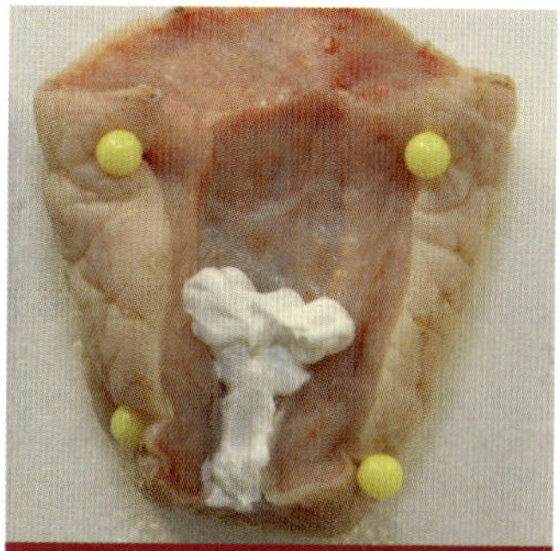

This shows how an internal teat seal fills the base of the teat and the teat canal to form a physical barrier helping to prevent infections

The teat seal forms an artificial plug or seal at the bottom of the teat and teat end if it is administered correctly. This seal remains in place throughout the dry period. Some of these teat seals have been proven to remain in place for up to 100 days.

You need scrupulous hygiene when administering a teat seal as it does not contain any antibiotic. When using a teat sealant on its own, the teat ends must be thoroughly disinfected before it is administered.

Internal teat seals

Administering the teat seal

It is very important that the teat is pinched at the top so that the teat is well sealed at the teat end. The teat seal will be infused up into the udder if the teat is not pinched and you will not get a good seal at the teat end. This will then be excreted from the udder for up to two or three weeks after calving. This can result in what is called black spot defect, or BSD for short, in cheese if these particles are not filtered out of the milk. Black spot defect is caused by particles of bismuth subnitrite in cheese.

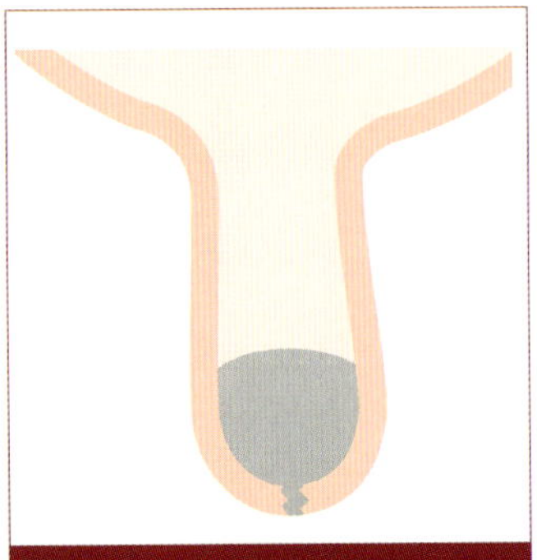

Correct - Seal at base of teat and in teat canal

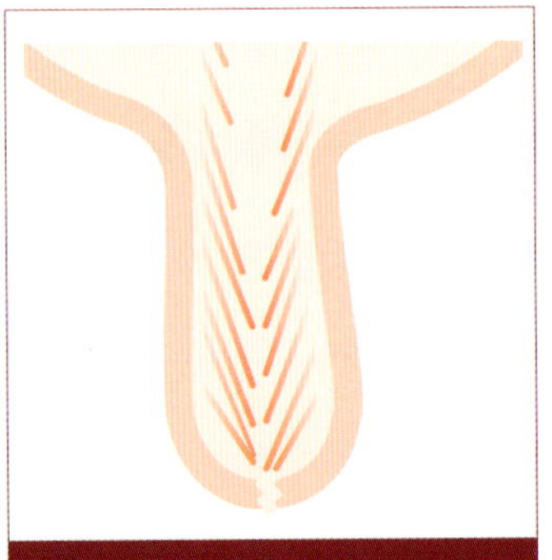

Incorrect - Teat not pinched before infusion. Seal pushed up into the teat and udder

TOP TIP

- Pinch the teat before administering teat seal

FACTS

- There is no milk or meat withdrawal when using a teat seal on its own
- Teat sealants have no adverse effects on calves

There is no milk or meat withdrawal if a teat seal is used on its own at dry off. When a teat seal is used in conjunction with a dry cow antibiotic, the withdrawal on the antibiotic dry cow tube must be followed.

There is a considerable amount of trial work that shows the reduction in clinical mastitis following the use of teat seals.

Quarters sealed at dry off;

- Were 30% less likely to develop a new intramammary infection in the dry period
- Showed a 33% reduction in clinical mastitis in the following lactation
- Were 83% less likely to have clinical mastitis due to environmental *Streps*

Internal teat seals

Which cows should be treated?

The majority of dairy farmers see a reduction of at least 25 – 30% in clinical mastitis and a reduction in the number of toxic coliform mastitis cases post calving.

You cannot predict which teats will remain open in the dry period and so teat seals should be used on ALL cows at dry off. You cannot prevent environmental bacteria from coming into contact with the teat end throughout the dry period and so there will always be a risk of dry period infections.

Several studies have shown that calves can ingest teat seal with no adverse effects whatsoever.

TOP TIPS

- Use teat seal on all cows at dry off
- One tube of teat seal per quarter at dry off
- Strip out the teat seal after calving

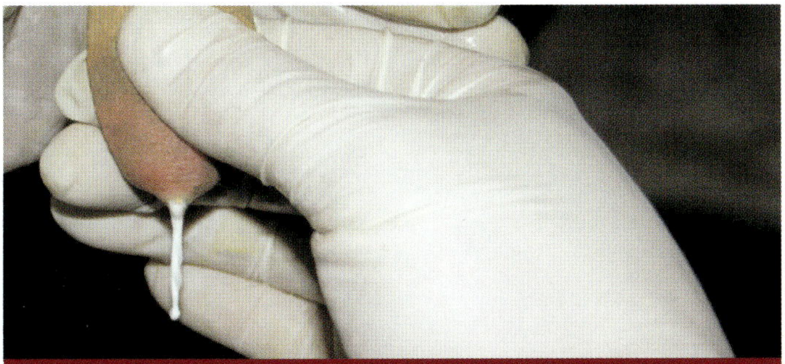

It is important to strip out the teat seal after calving

Teat seal debris on parlour floor

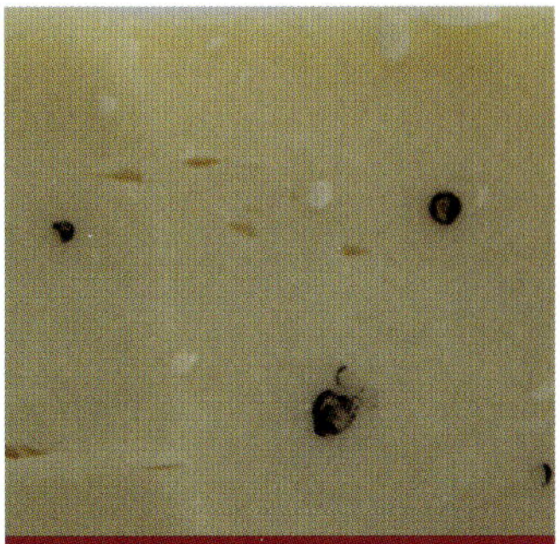

Black spot is a cosmetic problem of cheese, it has no adverse health risks to man or calves

Black spot is a cosmetic problem and has no impact on food safety whatsoever. Bismuth is frequently used to treat stomach disorders in humans. More and more dairy companies have installed clarifiers which will filter out any traces of teat seal to avoid this problem occurring.

Internal teat seals

Teat seal use in maiden heifers

A teat seal can be used on maiden heifers if they are likely to have problems with clinical mastitis around calving. It's not a procedure to recommend readily as there are significant risks to the operator and the heifer. However, mastitis problems around calving suggest that the environmental conditions where they are kept are poor. This should be the first area to be corrected.

Heifers which are kept and calve in clean conditions have minimal problems with mastitis from dry period infections. Maiden heifers are commonly treated with teat seals in New Zealand pre calving. This is because they calve their heifers and cows in very early spring in calving paddocks which are often overcrowded and dirty and at a time when the environmental conditions can be very challenging.

A small scale study in the UK showed that the incidence of mastitis in maiden heifers that were given an internal teat seal before calving was 1.8 cases per 100 animals in the first 30 days of lactation. This compared to 10.8 cases in the control group. This was a study of over 800 heifers at pasture in four seasonally calving herds that were randomly assigned to the different groups. Reductions were seen in each herd.

TOP TIP

- Keep all dry animals and those around calving in clean conditions

WATCH OUT!

- Take great care if using teat sealants in maiden heifers

	Control	Internal teat seal
A	9.2	0
B	22.8	2
C	6.5	2.2
D	11.8	2.07
Average	10.8	1.8

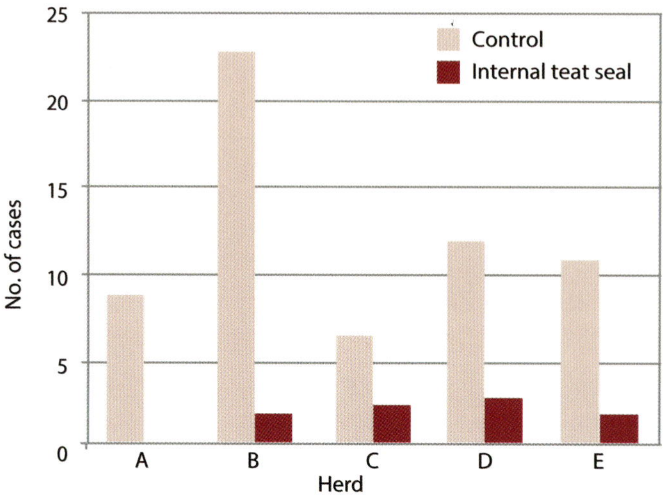

Changes in the number of mastitis cases in the first 30 days after calving in the control and teat sealant group

There can be significant risks to using a teat seal in heifers; both to the heifer and the operator. These heifers have never had their teats handled and so there is a real risk of the person infusing the teat seal being kicked and injured. For this reason the heifer should be carefully restrained by one person and the product infused by another in a secure and safe handling facility. When you infuse the teat seal tube you are breaking the seal at the base of the teat before calving and so there is a risk of introducing infections. It is essential that scrupulous hygiene is used at this time.

Cost benefit of using a teat seal

Comparison of two herds

There should be an economic return from using a teat seal. Let's compare the cost benefit between two herds.

Herd A

Let's take a herd of 100 cows with a mastitis rate of 50 cases per 100 cows per year. We expect a reduction of at least 25% in clinical cases, or 12 cases for this herd. The average cost of one case of clinical mastitis is around £250. So the saving in costs of mastitis cases is 12 cases x £250 which is £3,000 over the period of a year. If you save a cow with toxic E. coli mastitis which might have died then the saving is far more.

The cost of treatment will depend on the culling rate and the cost of the teat seal product. If this herd has a culling rate of 25% then 75 of the 100 cows will be dried off, the rest will be culled from the herd. Assume that the teat seal costs £8 per cow, so the total cost of teat seal for the herd is 75 cows x £8 or £600.

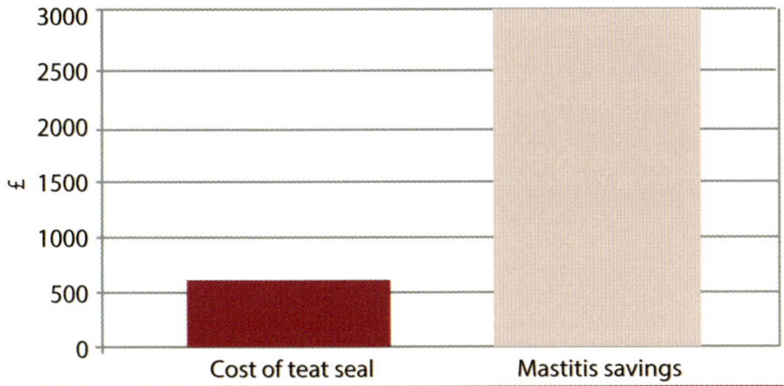

Herd A. Cost benefit from using a teat seal

This gives a cost benefit of over 500% for every pound, euro or dollar spent, you are getting five back!

KEY POINTS

- Cost/benefit of using a teat seal is very high even in herds with low levels of mastitis
- Dry cow therapy reduces cell count
- Internal teat seals prevent dry period infections

TOP TIPS

- Use teat seals in all animals at dry off
- Teat seals reduce clinical mastitis by an average of 25%
- They prevent new infections entering the udder during the dry period
- All farms should use internal teat seals

Cost benefit of using a teat seal

Comparison of two herds

Herd B

This herd also has 100 cows and has a mastitis rate of 30 cases per 100 cows per year which is the target rate of clinical mastitis. This herd has a culling rate of 20%.

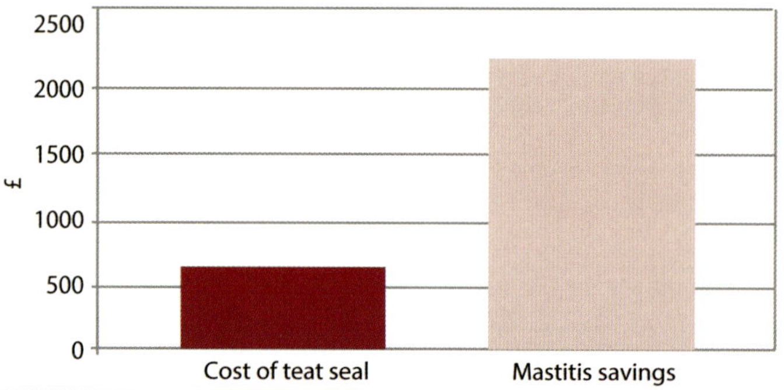

Herd B. Cost benefit from using a teat seal

You expect a reduction of at least 25% in clinical cases, or 8 cases for this herd, and at £250 per case this is a saving of £2,000. The cost of treatment will be 80 cows (20 will be culled without treatment) x £8 (for teat seal) which is £640. The cost benefit is still over 300% in this lower incidence herd with a low culling rate.

Breakeven point for use

For an average herd with a 25% cull rate, your breakeven point is a mastitis rate of 10! This means that if you use a teat seal when you have 10 cases of mastitis per 100 cows of year you cover your costs. If it's higher than this then you are in a payback situation! Of course, if there was a case of toxic *E. coli* mastitis the payback would be far greater.

These figures change according to the cost of each case of clinical mastitis, your culling rate and the cost of the internal teat seal. There are few veterinary medicines which will give such a good payback and the real benefit is that you see the reduction in mastitis.

Selective use of teat seals

What can go wrong?

The recommendation is for all cows to receive teat seal at dry off. The difficulty is that you cannot always predict environmental conditions during the dry period in the future. There will still be cows with open teats. If you save one case of toxic *E. coli* mastitis then you are likely to have more than paid for using teat seal for the whole year.

There are some herds which might want to only use teat seal on selected cows or at selected times of the year. Some might choose to use this only on high yielding cows, those with high milk flow rates or other at risk cows. High yielders and cows high milk flow rates will be more susceptible to dry period infections.

Some herds have a seasonal mastitis incidence. In exceptional circumstances where the environmental conditions are excellent and where the risk of dry period infections is minimal, you could stop using teat seal at this time. There are a number of pitfalls and this will be down to the individual or management rather than the product itself.

Risk - poor hygiene at infusion

This has resulted in the deaths of some cows shortly after dry off. This risk will be increased in cows which only get a teat seal if selective dry cow therapy is being used.

Solution

Use scrupulous hygiene at dry off to avoid this risk occurring. Follow the recommended best practice and you will have no problems.

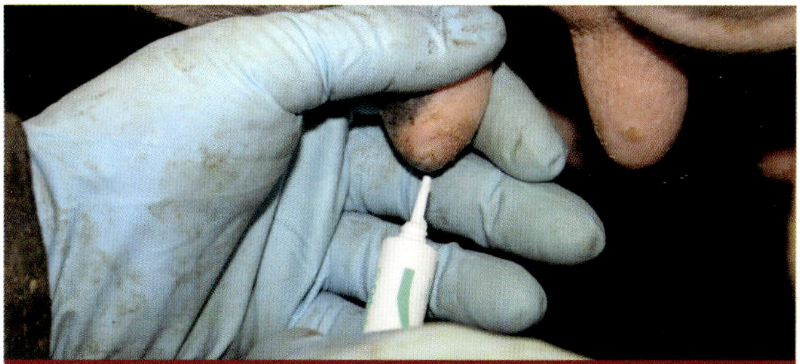

Expect mastitis problems if you infuse a teat seal without preparing the teat first

Risk - dirty dry cow environment

Teat seals will help overcome dry period infections but you still need a clean dry cow environment. Some herds think that when they use a teat seal they can overstock the dry cows or clean them out or bed them up less frequently.

Solution

The dry cow and calving environmental always has to be as clean as possible. You will have a few teats which might not be totally sealed and are still at risk following administration of a teat seal.

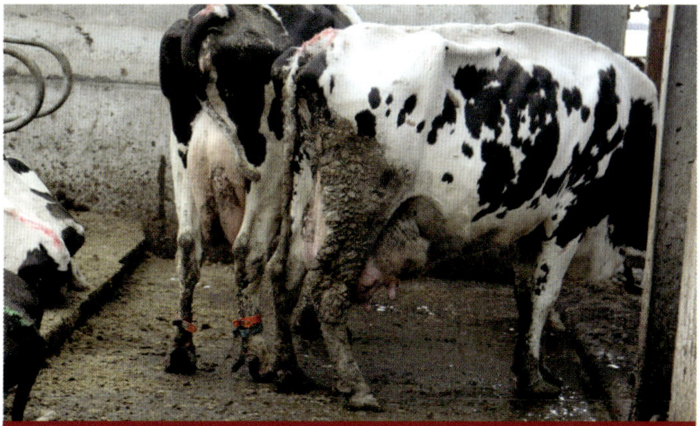

These dirty dry cows will still have a risk of dry period infections despite using a teat seal

Selective use of teat seals

What can go wrong?

Risk - warming teat seal tubes in water

The risk is that if this water is contaminated with bacteria such as *Pseudomonas,* then these will be infused into the udder on the nozzle of the tube. Cows can die from *Pseudomonas* infections introduced on the nozzle of dry cow or teat seal tubes.

Solution

Never place any intramammary tube in warm water to ease administration.

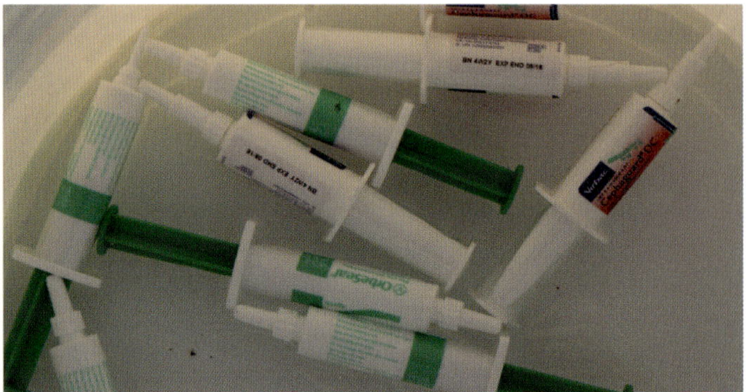

Tubes should never be put in warm water as there is a high risk of mastitis occurring

Risk - not pinching teats before infusing teat seal

This will infuse the teat seal up into the udder and so it's unlikely you will get a seal at the base of the teat. Cows will be more likely to milk leak when dry. You will also increase the chance of black spot defect if the milk goes to a cheesemaker who does not put the milk through a clarifier before processing.

Solution

Pinch the teat so that the teat seal remains at the base of the teat

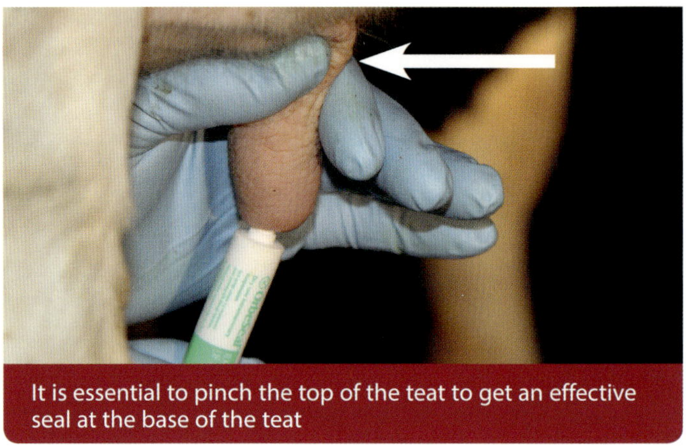

It is essential to pinch the top of the teat to get an effective seal at the base of the teat

Dry-off facilitator

Cabergoline

Cabergoline (Velactis, CEVA Animal Health) is a dry off facilitator that blocks prolactin, the hormone responsible for milk production. It is a new medicine that is injected within four hours of the last milking to assist with the dry off process.

Udder engorgement, discomfort and milk leakage are common problems seen after dry off. Various studies have shown that in excess of 30% of cows can leak milk after dry off increasing the risk of dry period infections.

Milk leakage can occur even if cows are given an internal teat seal at dry off. The greater the amount of udder engorgement, the greater the risk of milk leakage due to extra pressure on the teat canal. However, milk leakage can occur in all cows increasing the risk pf dry period infections.

One large scale study involving 900 cows on 63 farms in three European countries showed that cabergoline reduced milk leakage by 81% on the day after dry off. The study cows were giving 13 litres/day or more at dry off. None of the cows were treated with an internal or external teat sealant. This study also showed a 21% reduction in new intramammary infections in the dry period.

Cabergoline significantly reduces udder engorgement and reduces discomfort after dry off. A study of 228 cows on five farms was carried out to measure the effect of lying time and discomfort. On the day after dry off, treated cows were lying down for 798 minutes compared to 657 minutes for the control cows, an extra 2 hours 21 minutes! Treated cows were shown to be in less discomfort and so there are major welfare benefits from this treatment.

Capergoline allows farmers to milk cows twice a day without any changes to feed or milking frequency and then dry them off abruptly. This greatly assists with management, maximising production towards the end of lactation if drying off gradually, and helps to avoid negative energy balance and maximising immunity at this critical period.

Capergoline is not a replacement for an internal teat sealer, rather it will compliment their action and should reduce the risk of dry period infections further. This will have positive benefits on management, udder health and helps to maximise the welfare of the cow at dry off.

Milk leakage on a dry cow cubicle bed

Study herds

Mastitis incidence following use of teat seal

Here are three herd examples of changes to mastitis incidence following the use of a teat seal in the author's veterinary practice. Every herd which started using a teat seal has seen the benefits. No farmer has stopped using it at dry off.

Herd 1

Milks 250 cows and had a mastitis rate of 46 cases of mastitis per 100 cows per year. The herd made two changes - using teat seal at dry off in all cows and predipping to improve teat preparation.

With predipping, you would expect a reduction in mastitis across all cows throughout lactation. The results show clinical mastitis cases in the first 30 days of lactation showed a significant reduction with 21 cases of mastitis in the 10 months before the teat seal was used compared to 5 cases in the same period afterwards - a reduction of 75% and clearly shows this herd had major problems with dry period infections.

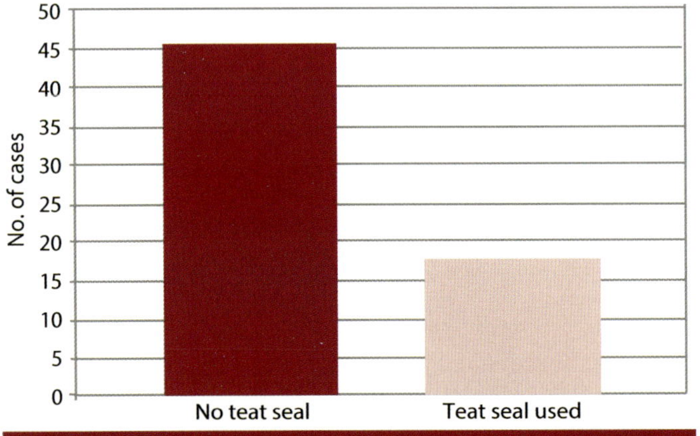

Herd 1 mastitis cases before and after teat seal use

The overall mastitis rate dropped from 46 to 18 cases per 100 cows per year, 60% reduction in total, but there was a greater reduction in cases within a month of calving and this was due to the benefits from using an internal teat seal.

The farmer was very pleased with the reduction in cases post calving and commented that it was now rare to get a toxic case of *E. coli* mastitis after calving. He estimated that he got over a 400% payback on his teat seal use.

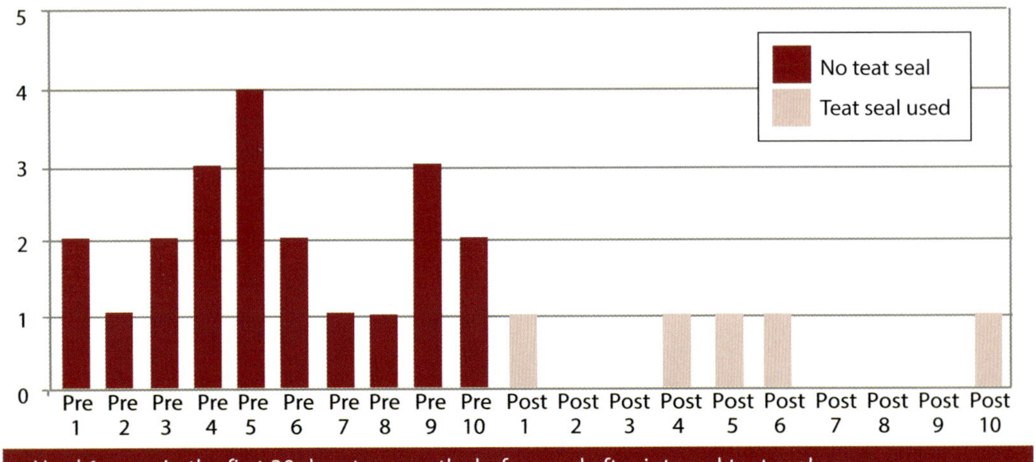

Herd 1 cases in the first 30 days ten months before and after internal teat seal use

Study herds

Mastitis incidence following use of teat seal

Herd 2

Milked 250 cows which were at pasture in the summer and housed on cubicles in the winter. The dry cows were housed in a straw yard during the winter months. The mastitis rate reduced from 57 to 29 cases per 100 cows per year. The farmer could only suggest that this reduction was due to the use of a teat seal as nothing else on the farm had changed.

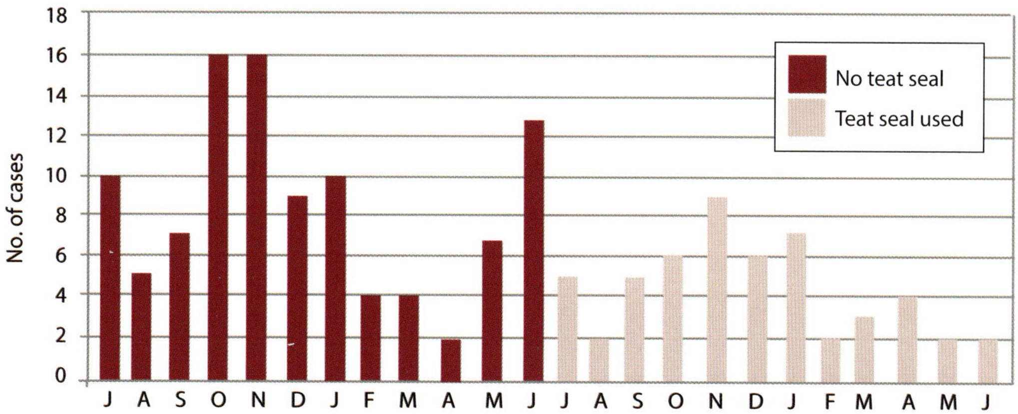

Herd 2 monthly mastitis cases for a year before and after teat seal use

You expect a 25-30% reduction in clinical mastitis following use of teat seal. A 50% reduction suggests that dry period infections were a significant problem causing clinical mastitis in this herd.

Herd 3

This herd had major problems with mastitis in the first month after calving with about one third of cows getting mastitis at this time. This is four times the target of 8%, one cow in twelve. This farmer started to use teat seals and at the same time reduced his stocking density in the calving pens.

He saw the number of cases in the first month post calving reduce from 16 in the year before to three in the year after internal teat seal use. He had no more toxic cases or deaths from *E coli* mastitis. These are exceptional reductions in mastitis within a month after calving and indicate very high levels of dry period infections.

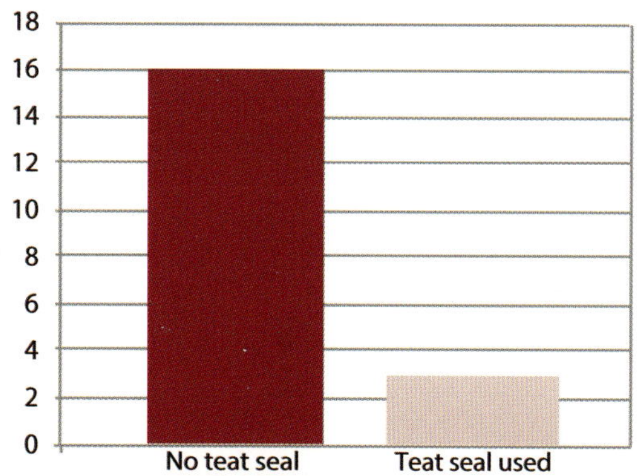

Herd 3 mastitis cases in the first month after calving for the year before and after internal teat seal use

How to administer a teat seal

Administer antibiotic dry cow therapy and/or an internal teat seal

Scrupulous hygiene during administration is essential. Preparation is everything and a small amount of extra time will pay dividends.

It is recommended that cows to be dried off are kept separately and brought into the parlour at the end of milking. This will allow plenty of time to administer treatments without the distraction of milking.

This protocol covers the administration of dry cow therapy and/or an internal teat seal. Amend the protocol if cows are only getting antibiotic dry cow therapy or an internal teat seal.

Best practice

1. Mark the cows to be dried off with a red spray or whatever other marking method you use on the farm. This clearly identities the cow before treatment is administered

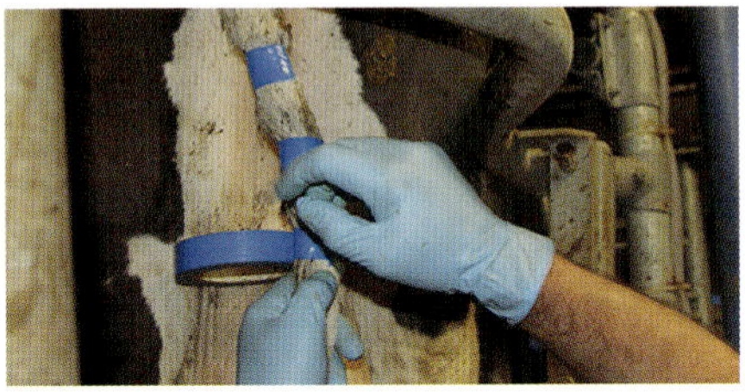

2. Wear clean gloves during the drying off procedure

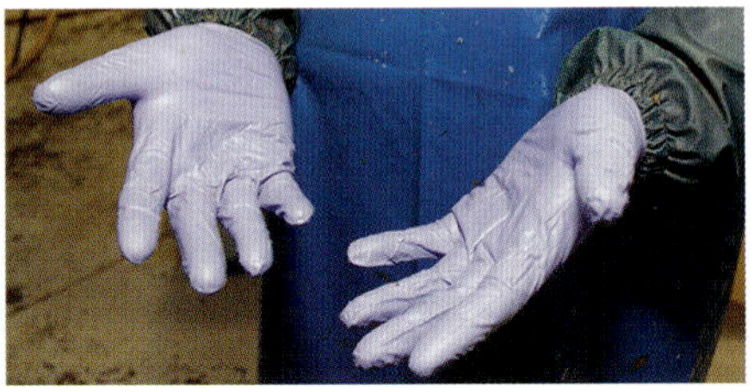

3. Strip out all quarters

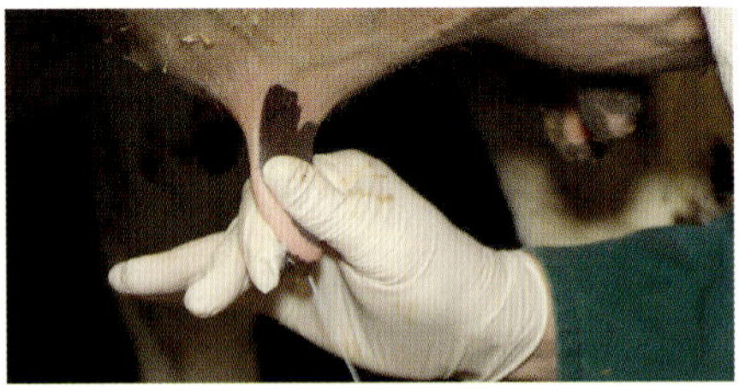

4. Predip the furthest two teats. If predip is not available, use a post dip solution. Allow a contact time of 30 seconds and wipe off this solution with a clean dry paper towel

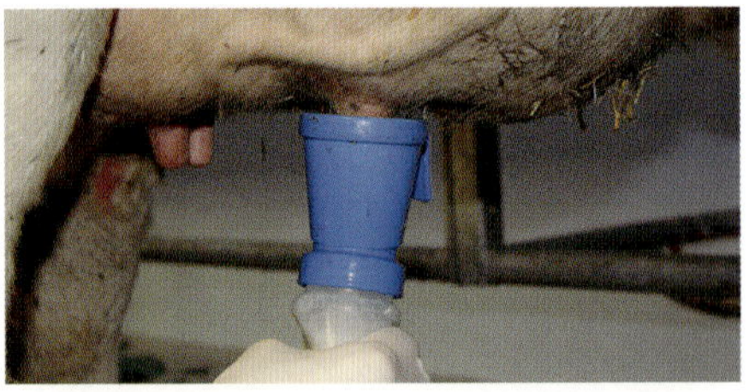

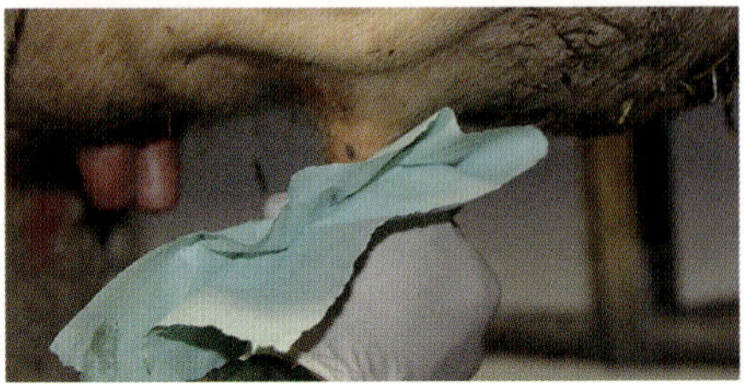

How to administer a teat seal

Administer antibiotic dry cow therapy and/or an internal teat seal

Carry out the procedure below on the two furthest teats first and then repeat on the two nearest teats or just individual teats. This is to ensure that there is no contamination of teat ends during this process.

5. Scrub the teat end with cotton wool and surgical spirit until no more dirt appears on the cotton wool. You can also use the medicated wipes provided with intramammary tubes. Repeat this task if teats are dirty

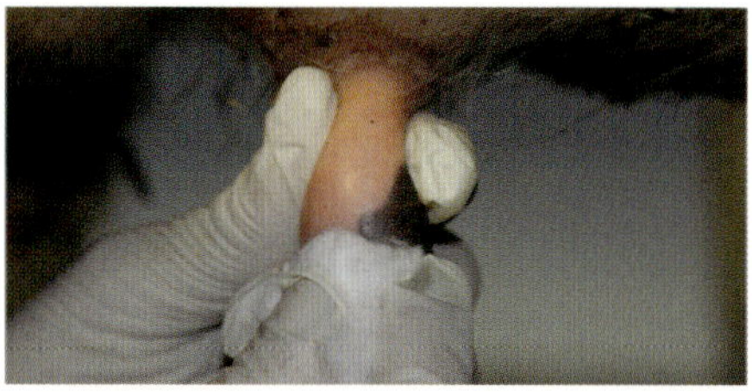

6. Infuse dry cow therapy if used, if not move to step 8

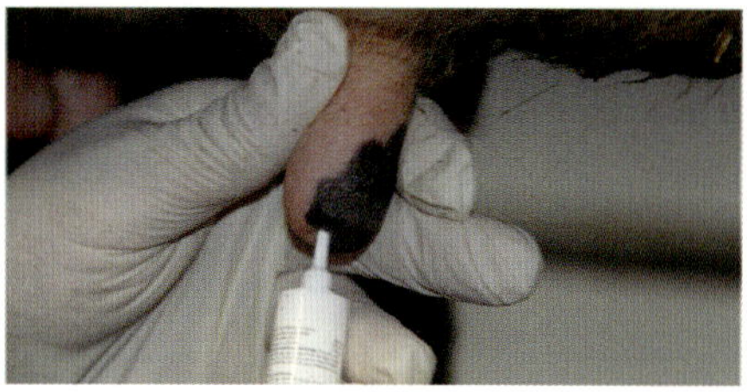

7. Massage up the teat

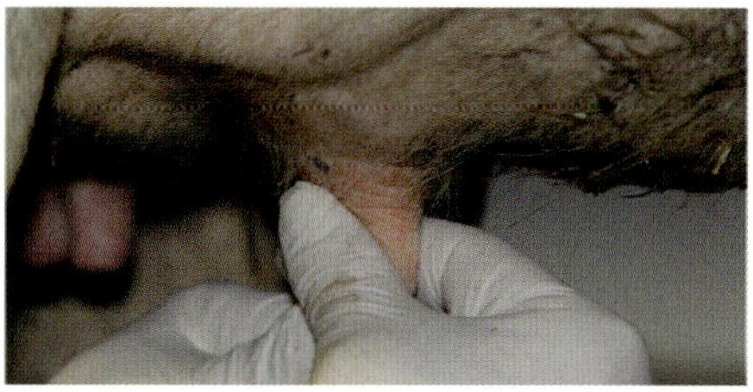

8. Pinch the teat at the base of the udder. Gently infuse the internal teat seal so that it remains at the bottom of the teat. **Do not massage up into the teat as it needs to remain at the base to form a physical barrier**
Repeat steps 5 to 8 on the two closest teats. Some people prefer to treat one teat at a time

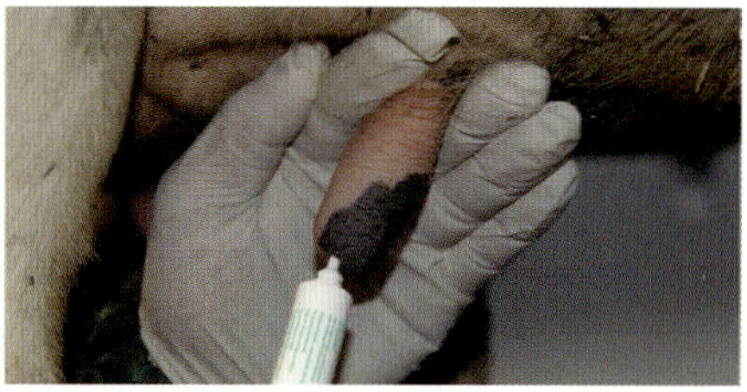

9. Teat dip all four teats with post dip solution

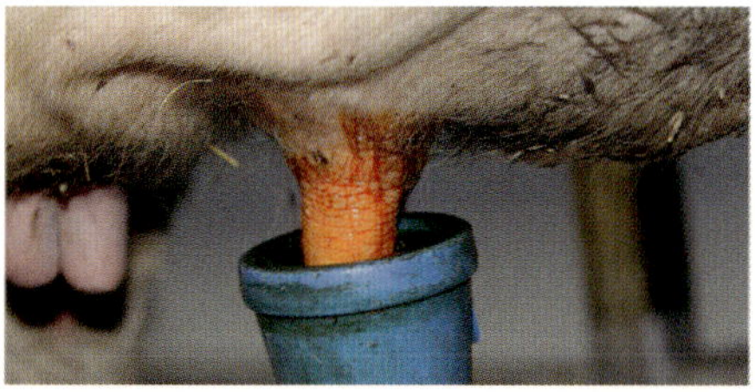

10. Record all the treatments into the medicine book

11. Cows should be put with the dry cow group

Other associated factors

How to control clinical mastitis

8

Nutrition

Factors which affect the immune system

Nutrition can affect mastitis in two key ways. Firstly if there is inadequate nutrition then this can affect the cows immune system and secondly, if the faecal consistency changes.

There are a number of nutritional factors which can influence mastitis. These include hypocalcaemia (commonly called milk fever), ketosis, Vitamin E or Selenium deficiency etc.

Hypocalcaemia

Hypocalcaemia or milk fever is where the blood level of calcium is low. This most commonly occurs after calving. Jersey and Guernsey animals are very prone to this condition. The animals are unable to get up and remain recumbent. The risk of mastitis will rise further if these cows end up lying in dirty conditions. Cows that are treated with calcium around calving have a 23 fold increased likelihood of developing *E. coli* mastitis. Cows which had an assisted calving will have an eleven fold likelihood of developing *E. coli* mastitis.

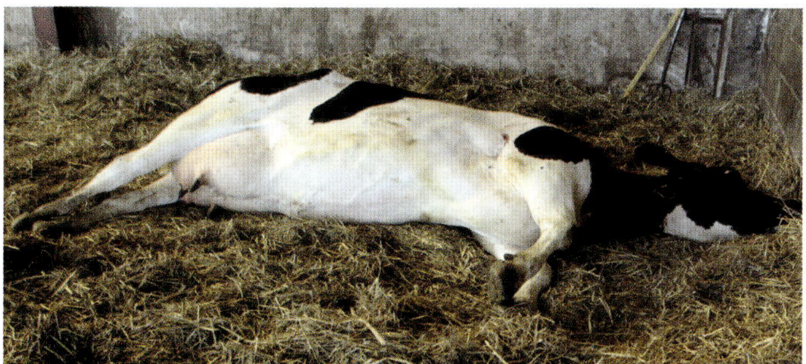

A cow with milk fever

FACTS

- A cow with milk fever has a 23x greater risk of *E. coli* mastitis
- Cows that had an assisted calving will have a 11x greater risk of *E. coli* mastitis

Nutrition

Factors which affect the immune system

Ketosis

Ketosis is where cows are in a negative energy balance. This most commonly occurs in early lactation and will be a greater risk in thin cows that are giving high amounts of milk. Cows with clinical and subclinical ketosis will be more prone to mastitis and other diseases. Ketosis levels can be monitored by blood and milk testing.

Ketosis can affect the ability of white blood cells to engulf and destroy bacteria and the other udder defence mechanisms.

In one Canadian study cows with ketosis had 50% more clinical mastitis in the first and second week of lactation compared to healthy cows.

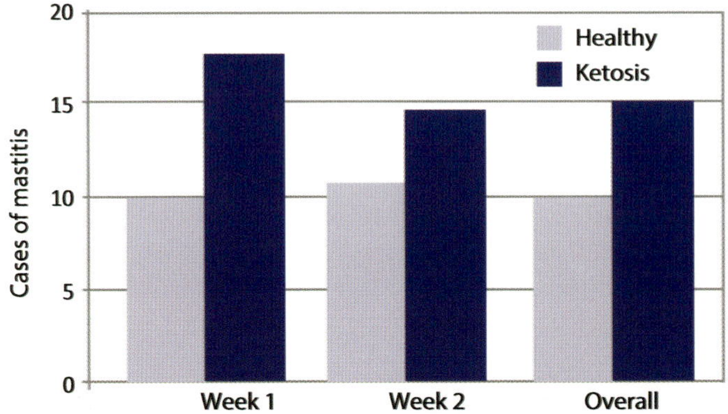

Proportion of healthy and Ketosis cows developing mastitis in the first two weeks of lactation

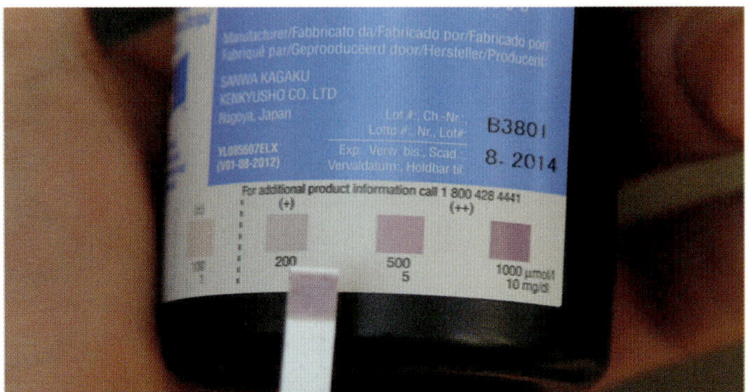

Ketotest milk dipsticks are a very quick way of measuring ketone bodies. The colour change here shows very high levels of Ketosis

FACTS

- Ketosis reduces the body's ability to fight disease
- Ketosis increases the risk of clinical mastitis

TOP TIP

- Monitor levels of ketosis by testing milk samples in cows after calving

Nutrition

Factors which affect the immune system

Vitamin E and selenium

Vitamin E and selenium are important for effective immune function. Levels can be low especially in herds that are feeding high levels of maize or corn silage. Vitamin E and selenium deficiencies can become a problem during the dry period when cows can have poor nutrition. Most of the work on mastitis and Vitamin E comes from the US.

A US clinical study showed that supplementing a deficient diet with vitamin E with or without a selenium injection reduced the incidence of clinical mastitis during the subsequent lactation by 37%.

FACT

- Levels of clinical mastitis will be higher if there is deficiency of Vitamin E or selenium

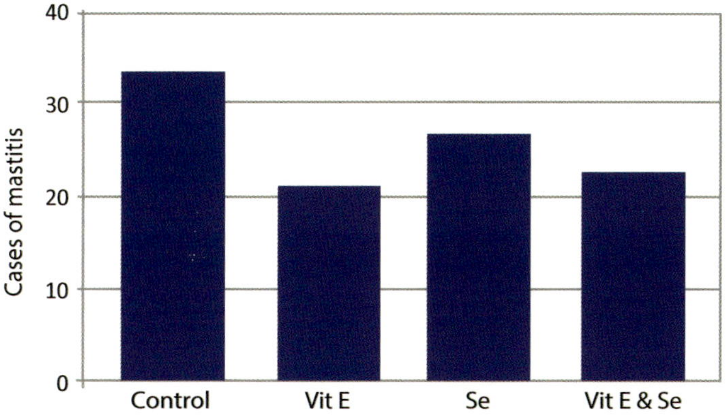

Clinical mastitis cases after supplementing deficiency with selenium (Se) and or Vitamin E

Vitamin E and selenium deficiencies are more common when high levels of maize silage are fed

Vitamin E and selenium status can be assessed by blood testing

Nutrition

Changes to faecal consistency

When the cow passes dung it should sound like a slow handclap when it hits the ground. It should form what is called a 'cow pat'. If the cow has loose faeces or diarrhoea then this will be spread far more in the environment that a traditional 'cow pat'. This increases the risk of the udder and teats becoming dirtier. This will be especially the case in straw yard systems where a greater area of straw will become contaminated with faeces. Loose faeces or diarrhoea is most likely going to be caused by nutritional problems.

FACT

- Cows with loose faeces will contaminate more of the environment and increase the risk of clinical mastitis

TOP TIP

- Watch faecal consistency to avoid environmental contamination

A good cow pat

Diarrhoea or loose faeces increase the risk of mastitis, especially in a straw yard

Machine milking

Servicing of the machine

It is essential that the milking machine is functioning correctly so that milk is removed from the udder swiftly with minimal impact to udder health. A faulty milking machine can affect mastitis in a number of ways, such as by damaging the teats or injecting bacteria into the udder.

It is important that the parlour is serviced at least twice a year with 2x milking and three times a year with 3x milking. The machine should be tested to approved standards by a trained engineer who leaves a written report listing the test results, faults that were corrected and items which require further attention. Any faults should be discussed with the farmer.

KEY POINT

- Service your milking machine at least twice a year with 2x milking

Always use a qualified engineer to test your parlour

Teat damage

Teat condition should be regularly assessed immediately after the milking machine is removed from the cow. If problems are identified then the milking machine engineer or dealer should be contacted to check the machine and to correct any faults contributing to damage.

Hyperkeratosis is one of the most common causes of teat damage that is seen. Remember that the teat canal is the primary defence mechanism to stop bacteria entering the udder. The risk of mastitis will increase if there is any damage to the teat or teat canal.

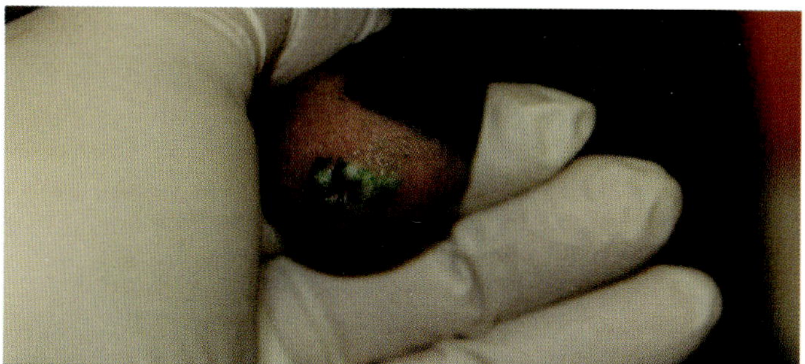

The risk of mastitis increases significantly with hyperkeratosis

FACT

- Levels of clinical mastitis will be higher with hyperkeratosis

TOP TIP

- Regularly assess teat condition immediately after the cluster comes off the cow

Machine milking

Liner changes

The liner is the only part of the machine which comes into direct contact with the teat. Liners have a limited life as they are constantly opening and closing during milking. If cows are milked with worn liners this can this can increase the risk of spread of infection from cow to cow, liner slip and teat damage through an increase in the unit on time and changes to the phases of pulsation.

In the UK, rubber liners should be changed every 2,500 milkings or twice a year, whichever is the shorter for 2x milking, and every four months or 2,500 milkings for 3x milkings. The time period is due to the corrosive effect of wash-up chemicals on the liners. You can work out the frequency of liner changes in the Appendix.

Liner usage can easily be checked by using the formula below. The number of cows refers to the average number of cows being milked. In this example, it is assumed that liners need to be changed every 2,500 milkings or six months, whichever is the lower figure, as cows are milked twice a day.

$$\text{Liner usage} = \frac{\text{Average no. of cows milking} \times 2^{*} \times \text{no. days between changes}}{\text{No. of milking units}}$$

*Change to 3 if milking three times a day.

For example, take a herd with an average of 160 cows milked twice a day through a parlour with 12 milking units where the liners are changed every six months (183 days).

$$\text{Liner usage} = \frac{160 \text{ cows} \times 2 \times 183 \text{ days}}{12 \text{ milking units}} = \textbf{4,880 milkings per liner}$$

In this case the liners have exceeded their useful life of 2,500 milkings. Always check the manufacturers recommendations for your liner life. They may need to be changed more frequently.

The frequency of change can be worked out by using the following formula:

$$\text{Liner life in days} = \frac{2{,}500^{*} \times \text{no. of milking units}}{\text{Average number of cows in milk} \times 2^{**}}$$

* Change if manufacturers recommendations are different.

** Change to 3 if milking three times a day.

$$\text{Liner life in days} = \frac{2{,}500 \text{ (milkings)} \times 12 \text{ (units)}}{160 \text{ (av. no. cows in milk)} \times 2} = \textbf{94 days or 3 months}$$

The life of silicone liners is longer and is worked out according to the elasticity remaining in the liner.

FACT

- Milking with worn liners will increase the risk of spreading mastitis bacteria from cow to cow, slow down milking and reduce milk yield

TOP TIP

- Ensure your liners are changed at the correct intervals

WATCH OUT!

- Liners need to be changed more frequently when you move from 2x to 3x milking

Regular liner changes are essential for mastitis control and efficient milking

Vaccination against E. coli mastitis

J5 E. coli vaccine

There are a number of mastitis vaccines effective against *E. coli* mastitis. One is called the J5 *E. coli* vaccine. Be aware that not all *E. coli* mastitis vaccines work in the same way or are as effective as each other.

Vaccines do not reduce all *E. coli* infections, but reduce the severity of infection. This increases the number of infections which undergo self-cure as shown in the diagram below.

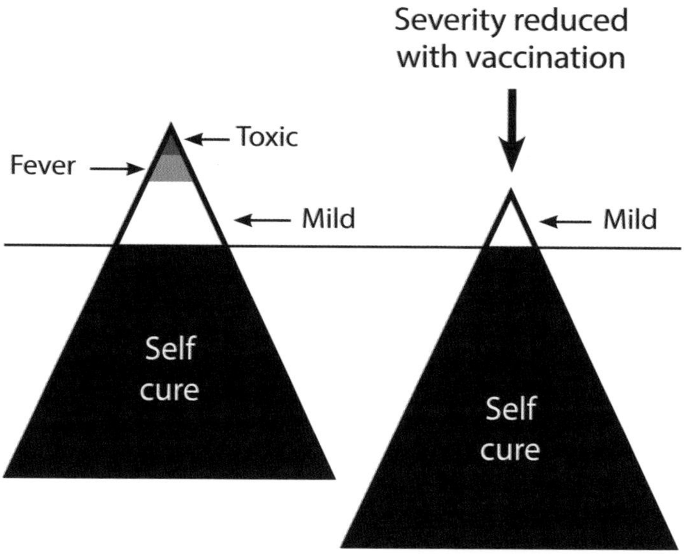

Effect of *E. coli* mastitis vaccination on severity of mastitis

E. coli strain J5 has a common antigen that is found in all strains of *E. coli*. It's a bit like a steering wheel of a car. There are lots of makes and models of cars but they all have one thing in common, a steering wheel. This is why this vaccine is effective against all the 300 or so strains of *E. coli*.

Vaccination against E. coli mastitis

Efficacy of vaccination

There are numerous studies which show the efficacy of the J5 mastitis vaccine against *E. coli*. In one US natural challenge study only 20% vaccinated cows developed clinical mastitis compared to 67% in unvaccinated controls.

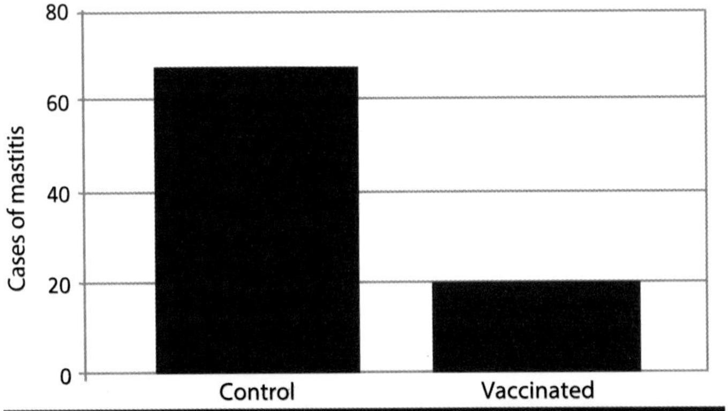

Cases of mastitis for vaccinated and control cows

In two further experimental challenge trials, all vaccinated cows developed clinical mastitis but the severity of mastitis was reduced. The vaccinated cows had lower peak numbers of *E. coli* and lower temperatures compared to cows which were not vaccinated. In addition, milk production and feed intake were more depressed in control cows compared to those which were vaccinated. A second experimental challenge trial showed that the vaccinated cows had a reduced duration of infection and reduced clinical signs.

Vaccination of cows with the J5 vaccine is very commonplace in the US with over 50% of cows vaccinated. The US vaccine is very competitively priced and offers a high cost/benefit ratio. It is essential that there is a good financial return from use of any vaccine. Internal teat sealants also reduce the incidence of *E. coli* and *Strep uberis* mastitis.

Some *E. coli* mastitis vaccines recommend that cows are vaccinated at dry off, before and after calving. This gives maximum protection around and after calving and at the peak of lactation where there is a high risk of infection.

FACT

- Vaccination against *E. coli* mastitis can reduce the severity of disease

Vaccination against E. coli mastitis

Benefits of vaccination

Some farmers prefer to use a rolling vaccination programme as it is easier to manage compared to following the data sheet recommendations. This is where animals receive their initial vaccination course so that the maximum levels of antibodies are produced and then move on to booster vaccinations every three or four months thereafter.

Benefits of *E. coli* mastitis vaccines

- ✔ Reduced severity of infection
- ✔ Reduced peak numbers of *E. coli* bacteria
- ✔ Fewer cows with fever
- ✔ Less reduction in milk yield
- ✔ Less feed intake depression
- ✔ Reduced duration of infection
- ✔ Reduced signs of mastitis

Top tips

How to reduce clinical mastitis

9

Top tips to minimise clinical mastitis

The successful approach to tackling clinical mastitis includes:

1. **Teats and udders should be clean.** Keep cows on clean and dry beds. If udders and teats are dirty, the beds are not clean enough. Remember to observe the cleanliness of your cows throughout the year.

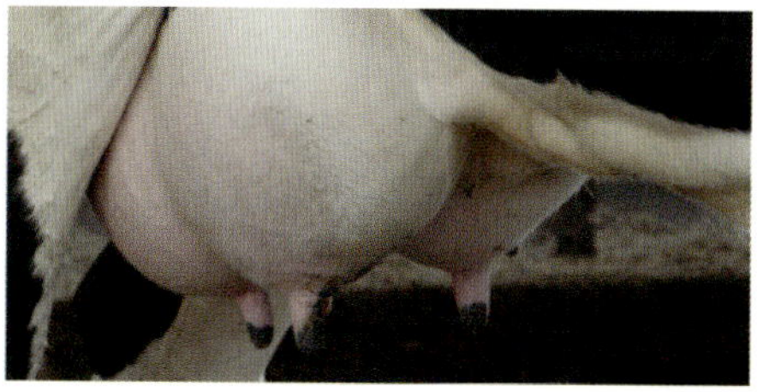

2. **Always use plenty of clean dry bedding.** Dry bedding absorbs moisture to keep beds dry. Make sure that passageways are scraped every time that cows go for milking.

3. **Calving pens.** These must be kept as clean as possible as freshly calved cows are most prone to toxic mastitis. Make sure that they are cleaned out frequently.

4. **Cubicles.** Ideally you should have 5-10% more cubicles than cows. Cows like choice where they lie. A heifer will not lie beside the dominant cow in the herd.

5. **Straw yards.** Allow adequate lying and loafing space according to the breed and size of cow. Bed up daily with plenty of clean dry straw and clean out every 2-3 weeks.

Top tips to minimise clinical mastitis

6. **Meticulous hygiene at all times during milking and predip cows.** Maintain meticulous cleanliness in the parlour at all times. Your milking parlour is a food factory. Predip to disinfect teats prior to milking. Teats must be dried after predipping. This is the gold standard. If the milk filter is dirty after milking you should review teat preparation.

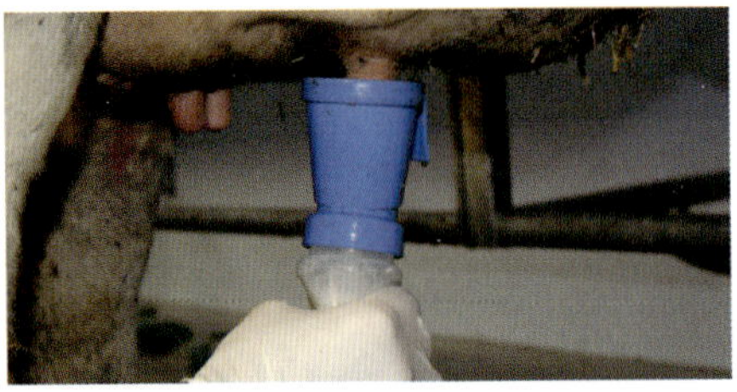

7. **Encourage cows to remain standing for 30 minutes** after every milking to give time for the teat canal to close. The best way is to do this is to always have fresh feed available after cows leave the parlour unless they are walking straight out to grass.

8. **Milking machine.** Make sure you have **stable vacuum levels** throughout milking. Make sure that it is serviced regularly and carry out daily maintenance. You should not have liner slip and vacuum levels should always be stable.

Top tips to minimise clinical mastitis

9. **Teat condition.** Regularly check the teats and if there is damage consult your milking machine engineer or manufacturer. There is an increased risk of mastitis if there is damage as the teat-end is the primary mastitis defence mechanism keeping bacteria out of the udder.

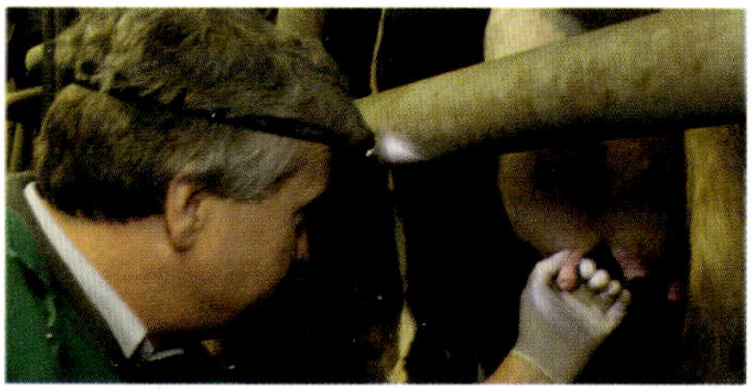

10. **Internal teat sealant.** This should be given to all cows at dry off especially if there are problems with mastitis in the first 30 days after calving. Internal teat seals reduce clinical mastitis levels by an average of 25 to 30%.

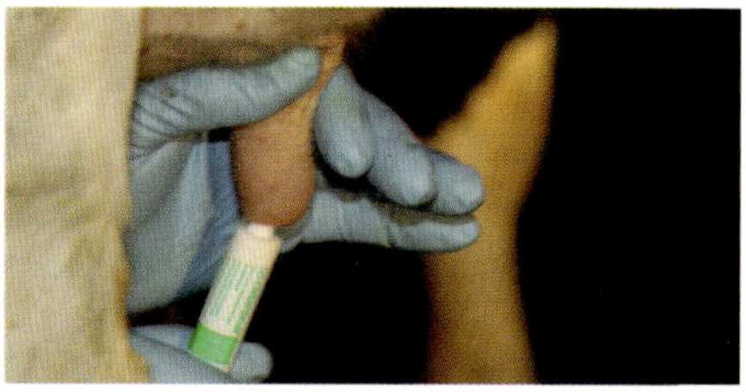

11. **Nutrition.** Make sure that cows are fed a balanced diet. Cows that have loose faeces or diarrhoea will contaminate the environment and increase the risk of clinical mastitis as the manure will splash everywhere contaminating udder and teats.

12. **Bacteriology sampling of clinical cases.** Identify the cause of mastitis in your herd and then fine tune your control measures.

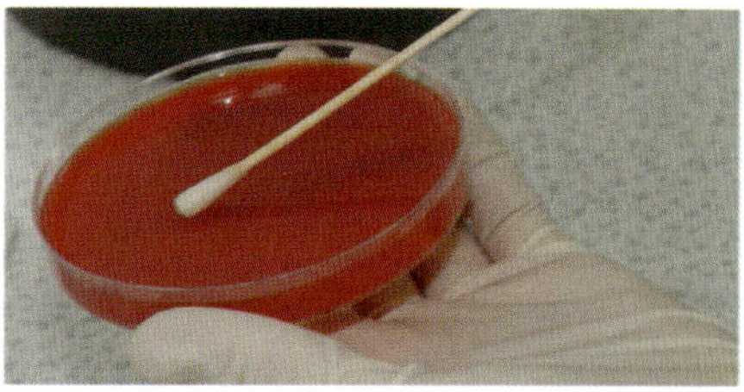

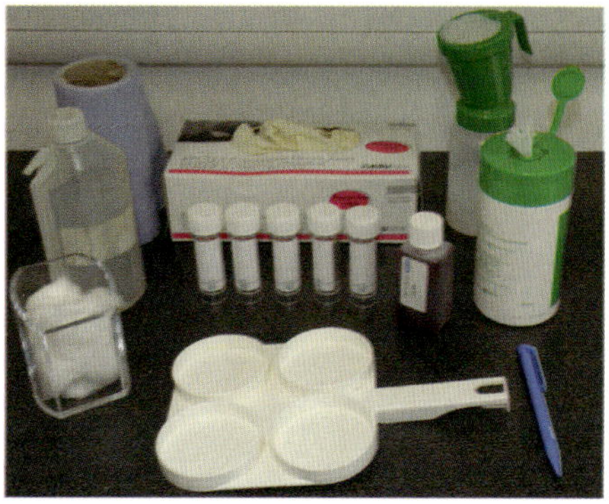

Top tips to minimise clinical mastitis

13. **Treatment.** Antibiotic therapy is appropriate for clinical mastitis due to *Staphs* and *Strep* infections but has no benefit for *E coli* and *coliform* mastitis in cows with normal temperatures.

14. **Records.** Keep accurate records of clinical mastitis including cow number, date, quarters infected, treatment, outcome and bacteriology results.

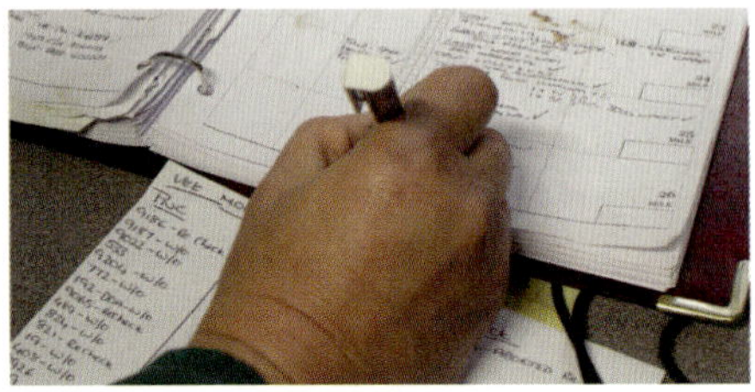

15. **Cull cows with chronic mastitis.** Cows that have had three or more cases in the same quarter or five cases in total throughout lactation should be culled.

16. **Vaccination.** Vaccination against *E coli* mastitis may be helpful if this is a significant problem.

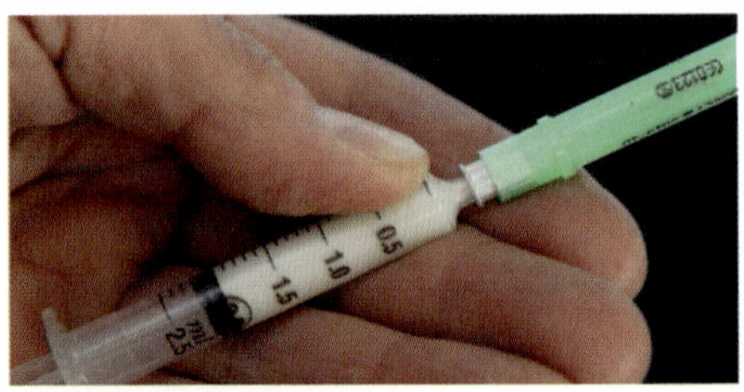

17. **Staph aureus**. If *Staph aureus* is a common cause of clinical mastitis then take steps to control this bacteria. These measures are covered in Mastitis Handbook 1. *'How to control somatic cell counts'.*

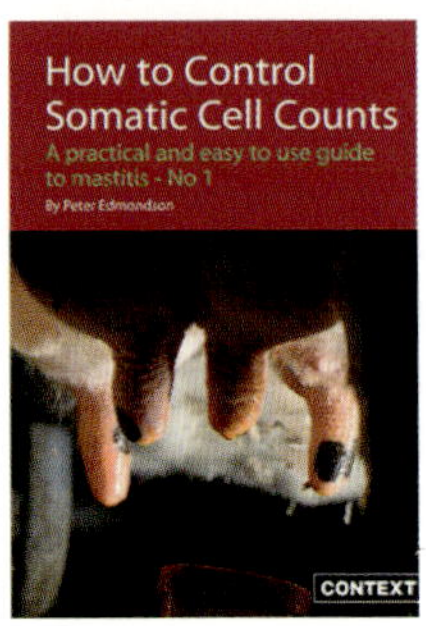

Appendix

How to control clinical mastitis

10

Liner change charts

Two times a day milking

Liner life in days according to average number of cows milked and parlour size and assuming a liner life of 2,500 milkings

2x

Av. no. of milking cows	No. of milking units							
	6	8	10	12	14	16	18	20
50	150	182	182	182	182	182	182	182
60	125	167	182	182	182	182	182	182
70	107	143	179	182	182	182	182	182
80	94	125	156	182	182	182	182	182
90	83	111	139	167	182	182	182	182
100	75	100	125	150	175	182	182	182
110	68	91	114	136	159	182	182	182
120	62	83	104	125	146	167	182	182
130	58	77	96	115	135	154	173	182
140	54	71	89	107	125	143	161	179
150	50	67	83	100	117	133	150	167
160	47	62	78	94	109	125	141	156
170	44	59	74	88	103	118	132	147
180	42	56	69	83	97	111	125	139
190	39	53	66	79	92	105	118	132
200	38	50	62	75	88	100	112	125
210	36	48	60	71	83	95	107	119
220	34	45	57	68	80	91	102	114
230	33	43	54	65	76	87	98	109
240	31	42	52	62	73	83	94	104
250	30	40	50	60	70	80	90	100
260	29	38	48	58	67	77	87	96
270	28	37	46	56	65	74	83	93
280	27	36	45	54	62	71	80	89
290	26	34	43	52	60	69	78	86
300	25	33	42	50	58	67	75	83

Liner change charts
Three times a day milking

Liner life in days according to average number of cows milked and parlour size and assuming a liner life of 2,500 milkings

3x

Av. no. of milking cows	No. of milking units							
	6	8	10	12	14	16	18	20
50	100	133	167	–	–	–	–	–
60	83	111	139	167	–	–	–	–
70	71	95	119	143	167	–	–	–
80	62	83	104	125	146	167	182	–
90	56	74	93	111	130	148	167	182
100	50	67	83	100	117	133	150	167
110	45	61	76	91	106	121	136	152
120	42	56	69	83	97	111	125	139
130	38	51	64	77	90	103	115	128
140	36	48	60	71	83	95	107	119
150	33	44	56	67	78	89	100	111
160	31	42	52	62	73	83	94	104
170	29	39	49	59	69	78	88	98
180	28	37	46	56	65	74	83	93
190	26	35	44	53	61	70	79	88
200	25	33	42	50	58	67	75	83
210	24	32	40	48	56	63	71	79
220	23	30	38	45	53	61	68	76
230	22	29	36	43	51	58	65	72
240	21	28	35	42	49	56	62	69
250	20	27	33	40	47	53	60	67
260	19	26	32	38	45	51	58	64
270	19	25	31	37	43	49	56	62
280	18	24	30	36	42	48	54	60
290	17	23	29	34	40	46	52	57
300	17	22	28	33	39	44	50	56

Glossary

How to control clinical mastitis

11

Abscess – A collection of pus which is surrounded by scar tissue. Antibiotics are unable to penetrate an abscess

ACR – Automatic cluster remover to prevent overmilking

Acute mastitis - Sudden onset of mastitis with swelling, pain, abnormal milk and reduced milk yield.

Anaerobic - Organisms which will not grow in the presence of oxygen

Antibiotic – A substance which kills or inhibits bacterial growth

Antibody – Proteins made by the immune system that help to remove infections such as bacteria and viruses

Bacillus - An environmental bacteria

Back flushing – An automated system for disinfecting liners between cows

Bacteria – Single celled organisms that can cause infectious diseases

Bactoscan – A rapid automated test to measure bacteria in milk. It counts dead and live bacteria.

Blanket dry cow therapy – Antibiotic dry cow therapy that is administered to all animals at the end of lactation irrespective of their mastitis infection status

Blitz therapy – An antibiotic intramammary treatment for subclinical *Strep agalactiae* infection during lactation

Casein – A protein in milk that is used for cheese making

Chronic mastitis – Udder infection that has been present for a long period of time. This can create damage within the udder, form scar tissue and results in reduced milk production

Clinical – Inflammation of the udder which is shown by visible abnormalities of the milk or the udder itself

Closed herd – A herd where no animals are purchased. All replacements are homebred

Cluster – The milking unit which is made up of a claw piece, liners, shells and short milk and air tubes

Cluster dipping – Clusters are dipped in a disinfectant solution during milking with the intention to reduce the risk of spreading infection from cow to cow

CMT – Abbreviation of the California mastitis test. A cow side test to give an indication of the level of subclinical mastitis

CNS – Coagulase negative Staphylococci

Coliform Count - The number of coliforms in a milliliter of milk

Coliforms - A group of Gram negative bacteria which include *E. coli, Klebsiella, Serratia* and *Enterobacter*

Contagious – Bacteria which live in the udder and on the teats and are spread from cow to cow during milking

Dry cow therapy – Antibiotics administered into the udder at the end of lactation to help eliminate subclinical mastitis

Dry period infection - An infection that enters the udder when cows are not lactating. These often cause mastitis in the following lactation

Dry wipe - Where teats are prepared before milking by wiping teats with a piece of paper towel

Dynamic milking machine test – A test carried out on the milking machine as cows are being milked

Environmental – Bacteria which live in the environment and spread onto the udder and teats between milking. Can enter the udder during the milking process and when the cow is dry

Epithelial cell – Cells within the udder which make up the milk producing tissue

Forestrip – Examination of the first few squirts of milk before milking. This is used to detect mastitis and stimulate let-down

Gangrenous mastitis – Mastitis where the blood supply to the affected quarter or part of the udder stops leading to death of parts or the whole quarter

Hyperkeratosis – Teat end damage where the teat canal is pulled out through the teat end or office. Mainly a milking machine induced problem.

Hypocalcaemia - A metabolic condition when the blood level of calcium is low

Glossary

Immune system – The body's defence system against disease

Infection – The presence of infectious organisms in the body

Inflammation – A natural response of white blood cells to try and eliminate infection

Internal teat sealant – A non-antibiotic viscous substance infused into the teat at dry off to prevent bacterial infections entering during the dry period

Intramammary – Within the udder and most commonly refers to treatments which are administered through the teat canal

Keratin – A waxy material produced by the cells lining the teat canal. This helps to reduce bacteria entering the teat canal

Ketosis - A metabolic condition when cows are in a negative energy balance

Klebsiella - An environmental bacteria what is commonly associated with wood products used for bedding

Lactation period infection - An infection that enters the udder when the cow is lactating

Lactose – a sugar present in milk

Lactoferrin - An enzyme that binds with iron that stops bacterial growth in the udder

Let down reflex – A natural process using the hormone oxytocin to squeeze milk out of the udder. The let-down reflex occurs after stimulation of the teats

Liner slip – Occurs when the liner slides down the teat and allows air to enter the milking cluster. This can be due to a variety of factors including milking with wet teats, low vacuum levels, poor design of liners etc

Lipase – An enzyme that breaks down fat in milk

Mastitis – Inflammation of the udder caused by bacterial infections

Mastitis detector - A mesh filter situated in the long milk tube close to the cluster to help pick up clots to aid mastitis detection

Mastitis rate - A measure of the incidence of mastitis. Cases per 100 cows per year

Medicated towels - Towels that have been impregnated with a disinfectant solution to help kill bacteria on the teat before milking

Milk filter or milk sock - A filter to remove any traces of debris such as straw, sand etc.

Mycoplasma - An anaerobic contagious mastitis bacteria

Neutrophil – One form of white blood cell that engulfs bacteria

Oxytocin – The hormone responsible for milk let-down

Partial blitz therapy – Blitz therapy which is carried out on selected cows during lactation

Pasteurisation – Heat treatment of milk to kill bacteria which can spoil milk and also affect humans. Tuberculosis bacteria are killed through pasteurisation

Plasmin – An enzyme that breaks down milk protein and can carry on acting after pasteurisation

Post milking teat disinfection - A solution applied to teats after milking to help kill bacteria that have been transferred onto the teat during the milking process and to optimize teat skin condition

Predipping or premilking teat disinfection - A solution applied to teats before milking to help kill bacteria present when the cow comes into the parlour. This solution is always wiped off the teat before unit attachment.

Prototheca - An algae associated with contaminated water supplies

Proteus - An environmental organism

Pseudomonas - An environmental bacteria causing mastitis that is commonly associated with contaminated water sources

Pulsation – The opening and closing of liners during milking which results in milk flowing from the udder and the teat filling up with milk

Recurrence rate - The percent of quarters that have more than one case of mastitis

RMS - Recycled manure solids, a form of bedding

Scar tissue – Fibrous tissue that is created after infection which permanently replaces tissue. In the udder this can stop intramammary antibiotics distributing evenly throughout the udder

SCC – Abbreviation for somatic cell count

Selective dry cow therapy – Antibiotic dry cow therapy that is only administered to selected cows

Somatic cell – A body cell that enters the udder in response to inflammation. The majority will be white blood cells combined with a small percentage of epithelial cells from the milk producing tissues

Somatic cell count – The number of somatic cells in milk. This is measured per millilitre of milk and results of 200,000 are often referred to as 200

Staphylococci – One form of contagious mastitis that grow in grape like clusters. Staphylococci are divided into *Staph aureus* and CNS which are all the staphylococci apart from *Staph aureus*

Streptococci – Forms of contagious mastitis that grow in chains. The most common are *Strep agalactiae, dysgalactiae* and *uberis. Strep uberis* also has some environmental properties

Subclinical – Infection which can only be detected by testing. The most common form of mastitis and can result in significant economic losses

TBC – Total bacteria count to measure bacteria in milk. Bacteria are grown on culture media and it takes three days to get a result.

Teat canal – The passageway from the teat end into the udder which is normally 6 to 10mm long. There is a sphincter muscle which helps to ensure that the teat canal remains closed between milkings and during the dry period

Teat cup shell – The metal or plastic shell that surrounds the liner

Teat seal or internal teat seal - A paste infused at dry off to form a physical barrier at the base of the teat to help prevent dry period infections. It does not contain any antibiotic.

Toxaemia - A condition caused by toxins or poisons from an acute mastitis, such as *E. coli,* that enters the bloodstream and causes illness

Toxic mastitis - Severe mastitis where the cow is very ill and may be recumbent. It often has a sudden onset with pain, swelling of the udder, abnormal milk and milk production often stops.

Vaccine - A substance used to provide protection against a disease

Vacuum reserve – Vacuum produced over and above that required to run the milking machine to ensure that there is stability at the teat end throughout milking

Wet test – A test carried out on the milking machine as cows are being milked

White blood cell – Cells present in the blood and milk to kill bacteria

Yeasts - A range of environmental organisms

Index

How to control clinical mastitis

12